The Fracture of Brittle Materials

The Fracture of Brittle Materials

Testing and Analysis

Second Edition

Stephen W. Freiman
John J. Mecholsky, Jr.

Published by John Wiley & Sons, Inc., Hoboken, New Jersey.
Published simultaneously in Canada.

Edition History
John Wiley & Sons, Inc. (1e, 2012)

Registered Office
John Wiley & Sons, Inc., 111 River Street, Hoboken, NJ 07030, USA

Editorial Office
111 River Street, Hoboken, NJ 07030, USA

For details of our global editorial offices, customer services, and more information about Wiley products visit us at www.wiley.com.

Wiley also publishes its books in a variety of electronic formats and by print-on-demand. Some content that appears in standard print versions of this book may not be available in other formats.

Library of Congress Cataloging-in-Publication Data

Names: Freiman, S. W., author. | Mecholsky, John J., Jr., author.
Title: The fracture of brittle materials : testing and analysis / Stephen W. Freiman and John J. Mecholsky, Jr.
Description: Second edition. | Hoboken, New Jersey : John Wiley & Sons, Inc., [2018] | Includes bibliographical references and index. |
Identifiers: LCCN 2018036245 (print) | LCCN 2018036557 (ebook) | ISBN 9781118769669 (Adobe PDF) | ISBN 9781118769775 (ePub) | ISBN 9781118769706 (hardcover)
Subjects: LCSH: Fracture mechanics. | Brittleness.
Classification: LCC TA409 (ebook) | LCC TA409 .F765 2018 (print) | DDC 620.1/126–dc23
LC record available at https://lccn.loc.gov/2018036245

Cover design by Wiley
Cover image: Courtesy of Gina and Jason Blume

We want to thank Jason and Gina Blume for the picture of the natural fracture origin on the cover of this book.

Set in 11/13.5pts Times by SPi Global, Pondicherry, India

Printed in the United States of America

V10006424_112818

We want to dedicate this book to all the people who inspired us, guided us, and collaborated with us over the years. We especially want to acknowledge the influence of Roy Rice, George. D Quinn, and Shelly Wiederhorn in formulating many of our ideas. In many cases, they provided the necessary catalyst to further our knowledge through discussions and collaborative research.

Contents

Preface ix

Acknowledgements xi

1. Introduction 1

2. Fracture Mechanics Background 7

3. Environmentally Enhanced Crack Growth 23

4. Fracture Mechanics Tests 37

5. Strength Testing 79

6. Thermally Induced Fracture 129

7. Modeling of Brittle Fracture 145

8. Quantitative Fractography 167

9. Microstructural Effects 207

10. Reliability and Time-dependent Fracture 223

11. Concluding Remarks 235

Subject Index 239

Preface

The purpose of this book is to bring together the background, testing procedures, and analysis methods needed to design and use materials that fail in a brittle manner, primarily ceramics. In this context we define ceramics quite broadly as any inorganic nonmetal. Such a definition includes diverse materials such as semiconductors (e.g. Si, GaAs, InP), other single crystals (ZnSe, CaF, etc.), cements and concrete, and of course the oxides, carbides, nitrides, and others that we normally think of as ceramics. Ceramics are also used in composite form, either by dispersing one phase in another or by crystallizing phases from a glassy matrix. Most test procedures designed for monolithic bodies can be used here as well. However, continuous fiber-reinforced composites behave quite differently and will not be discussed herein. Ceramics are also increasingly used in films and coatings, but determining the mechanical properties of materials in these forms is more complex and will not be addressed in this book.

This book addresses testing and analysis at temperatures for which the material behaves in a brittle manner. At elevated temperatures other modes of failure often are important. These include creep as well as general plastic deformation. Both of these topics are outside the scope of this book.

In this book we provide the reader some of the background needed to understand the brittle fracture process as well as a basis for choosing the proper test procedures. The mathematical development of the expressions used to calculate the various properties will be kept to a minimum; the reader will be referred to fundamental references. We intend to provide examples to allow the reader unfamiliar with the tests to be able to perform the test procedures properly. However, the reader is strongly encouraged to consult formal national and international standards for more extensive test procedures. Questions to test comprehension for self-evaluation are given at the conclusion of each chapter.

Chapter 1 is a general introduction to the concept of brittle failure. Chapter 2 provides a condensed background into the basic principles of fracture mechanics that underlies most of the test and analysis procedures. Linear elastic fracture mechanics (LEFM) is the basis for measuring the fracture toughness of materials. Chapter 3 gives some background into the theory and mechanisms of environmentally enhanced crack growth, a process that is particularly important for designing components that are intended to survive over long periods of time under stress. Chapter 4 provides extensive details on fracture mechanics tests used to determine both a material's resistance to fast fracture and the parameters associated with environmentally enhanced crack growth. Chapter 5 addresses the test and analysis methods to determine the strength of ceramics. New in this chapter is a section describing test procedures applicable to biomaterials. Also new is a new procedure for determining the probability of failure within a set of ceramic components based on modern statistical concepts. Chapter 6 is a new addition that provides information on the causes of thermal shock failure, a common occurrence in ceramics. Also included are some of the test procedures that are used to rank the thermal shock resistance of such materials. Chapter 7 is also new; it describes attempts to model the fracture process and to provide predictions of resistance to crack growth. Chapter 8 provides a background and discusses the methods of understanding the fracture process based on quantitative measurements made on the fracture surface. Chapter 9 discusses an important topic with respect to polycrystalline materials, namely, the effect of the microstructure of the specific material. Chapter 10 provides the background, test methods, and analytical procedures needed to confidently predict the safe lifetimes of brittle components under stress. Finally, Chapter 11 summarizes the critical issues with respect to brittle fracture.

STEPHEN W. FREIMAN
JOHN J. MECHOLSKY, JR.

Acknowledgements

We gratefully acknowledge the work of Nicholas Mecholsky in preparing many of the illustrations in this volume. We also appreciate the numerous technical discussions with George Quinn and Jeffrey Fong. Finally, we would like to express our gratitude to Roy Rice for introducing us to many of the topics discussed in this book, particularly those focused on the effects of microstructure.

S. W. F.
J. J. M.

CHAPTER 1

Introduction

The properties of ceramics have made them extremely attractive to society in uses such as electrical and thermal insulators, high temperature crucibles for steel fabrication, elegant dinnerware, etc. More recently, their applications have become even more extensive and sophisticated, ranging from complex electronic devices such as multilayer capacitors and ultrasonic transducers to thermal protection for aircraft engines and applications in the dental and medical fields. However, the brittleness of ceramics, making them subject to sudden failure without prior warning, has at times limited more extensive use. Everyone knows that traditional ceramics, such as dishes and glasses, are brittle: drop a teacup or a plate, break a window, and you experience the brittleness. By brittle we mean that there are no mechanisms to relieve or alter the high stresses at crack tips, such as dislocations in metals or crazing in polymers. The lack of any stress relief mechanism results in cracks growing to failure at stresses that are significantly less than those necessary to initiate and propagate cracks in metals.

Despite their brittleness, advanced technical ceramics form the basis for a wide variety of important products. They are used in applications in which they experience significant stresses imposed by not only

The Fracture of Brittle Materials: Testing and Analysis, Second Edition.
Stephen W. Freiman and John J. Mecholsky, Jr.

mechanical loading but also thermal, magnetic, or electronic conditions. One sees ceramics everywhere: the large electrical insulators on poles, spark plugs, and skyscraper windows that must resist high winds. Some we do not see or are not aware of. Cell phones would not operate without ceramics having special dielectric properties; automobiles contain hundreds of multilayer ceramic capacitors. Aircraft engines depend on ceramic coatings to reduce the temperature of the metal blades. Turbine engines for auxiliary power generation are now being constructed with rotating ceramic blades.

Another use of ceramics that requires complete reliability is aluminum or zirconium oxide hip and knee replacements in the human body. Dental ceramic prosthetic composites are routinely implanted in many patients. The hardness, inertness, and wear resistance of these materials make them ideal candidates to replace metals in such situations. Particularly when the patient is young, the lesser amount of wear debris produced by the ceramic means that the component can be used in the body for a significantly longer time than one made of metal.

The list of ceramic applications is extensive, including materials that we do not normally think of as ceramics, e.g. semiconducting materials, such as silicon, gallium arsenide, etc., and oxide films crucial for the operation of electronic devices. Because of the brittleness of these materials and their similarity in mechanical behavior to conventional ceramics, we refer to each of these materials as *ceramics*. Figure 1.1 shows some prime examples of advanced technical ceramics.

In each of these examples and in the myriad other applications, the brittleness of ceramics necessitates that special care must be taken in determining the mechanical properties of the material and discovering the stresses imposed on the final product during operation. The fact is that unseen, and probably undetectable, defects can lead to catastrophic failure. We will call these defects *flaws*. By a flaw we do not necessarily mean that errors were made in production. While improper processing can lead to pores or inclusions, component failures caused by these are relatively rare. For the vast majority of the time, brittle failure begins at the surface of a component from small cracks that are produced during the machining, finishing, or handling processes. All ceramics contain such flaws; there is no perfect brittle material. Even the strongest ceramic, pristine glass fibers, contains small flaws in its surface despite

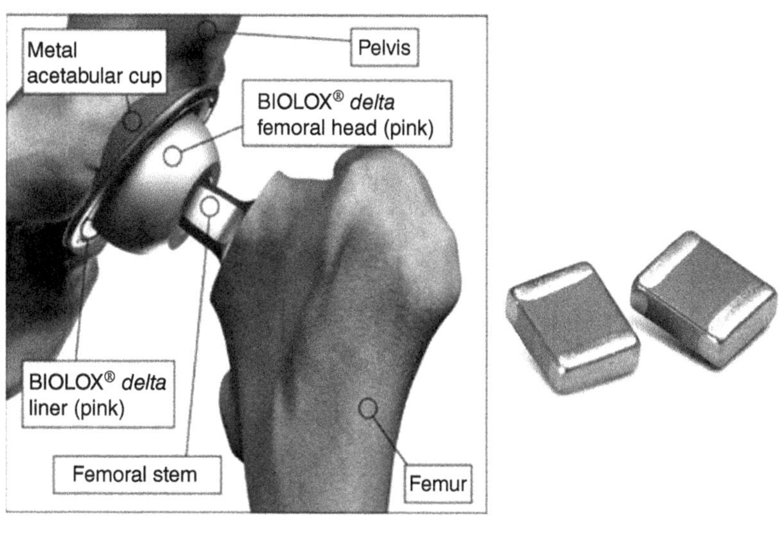

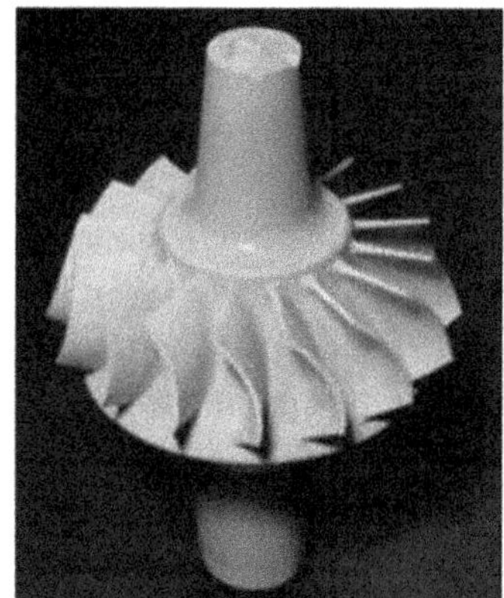

A 16-blade silicon nitride turbine wheel for use in small turbogenerators

Figure 1.1 Examples of advanced technical ceramics. From the left to right are an example of a ceramic hip replacement, barium titanate capacitors, various silicon nitride components, and a silicon nitride turbine wheel.

the care taken to avoid any surface damage. It is the size and shape of such flaws, i.e. the *flaw severity*, and their location with respect to the tensile stresses that determine the strength of a component.

Brittle fracture is a statistical process. We usually think of such failure in terms of a "weakest link" model. That is, failure begins from the most severe flaw located in the region of highest tensile stress. Also, the size of flaws in real components, 10–200 μm, means that detection of such defects by some nondestructive means prior to putting the part into service is extremely unlikely.

Another important aspect of most ceramic materials is that even if their strength when placed into service is sufficiently high that failure should not occur, in the presence of certain environments, e.g. water or water vapor, surface cracks will grow under the operational stresses, and failure can occur after a period of days, weeks, or even months. Fortunately, we have sufficient knowledge of this behavior, so that with proper testing and analysis, excellent predictions of the safe operating envelope, stress, and time can be given. Nonetheless, the user of ceramic components should recognize that such analysis only pertains to flaws that existed prior to putting the component in service. Other defects can be created during operation, e.g. from dust or rain, which may limit useful service life.

Knowledge of the brittle fracture process, most of which has been acquired over the past 30–40 years, has played a major part in our ability to design and use these materials, even in situations where the component is subject to significant tensile stresses. Two developments, which at the time were outside the field of materials science, were of major importance in contributing to our ability to safely use these materials. One was the development of the field of linear elastic fracture mechanics. Fracture mechanics provides the framework by which the effect of the stresses imposed on a body can be translated into predictions of the propensity of any cracks or flaws within the body to grow. This has led to the development of test methods and data analysis that permit designers to choose a material, machine it to shape without producing damage that could lead to premature failures, and carry out quality control procedures that provide confidence in the reliability of the part under operating conditions. A second important advancement, allowing us to design with brittle materials, was the development of statistical techniques that account for the uncertainties in the experimental measurements of the various parameters needed to make predictions of reliability.

A third factor that has greatly benefited the use of brittle ceramics in a wide variety of applications is the agreed-upon use of a common test methodology through national, regional, and international standards. Most of these standards have been developed by consensus by private standards development organizations such as ASTM International and the International Organization for Standardization (ISO). The details of the standards coming out of the deliberation process are based on years of data obtained in laboratories throughout the world.

In this second edition of the book, we summarize the concepts behind the selection of a test procedure for fracture toughness and strength determination and go into some detail in how the statistics of fracture can be used to assure reliability. We explain the importance of the role of microstructure in these determinations and emphasize the use of fractographic analysis as an important tool in understanding why a part failed. We have included a significant quantity of material related to the fracture of biomaterials. We have also included new chapters, one devoted to thermal shock and the other to the modeling of the fracture process. In addition, the portion of the book discussing how to treat the statistics of fracture strength data to ensure reliability has been greatly expanded.

CHAPTER 2

Fracture Mechanics Background

INTRODUCTION

At the most fundamental level, brittle fracture occurs when stresses reach the level needed to break the bonds between the atoms in the material. However, if there were no means of concentrating stress, loads necessary to cause failure would be extremely large. It is the presence of small defects that concentrate applied stresses to a magnitude sufficient to cause them to grow. The materials we discuss in this book are brittle because plastic flow mechanisms in them are insufficient to relieve the stress concentration at the defect.

The science of fracture mechanics allows us to calculate the forces needed to cause defects to grow based on knowledge of specimen geometries and the applied loads. In this chapter we provide some historical perspective and a summary of the basic principles of fracture mechanics. The reader is encouraged to consult Anderson (1995) and Munz and Fett (1999) for more details.

The Fracture of Brittle Materials: Testing and Analysis, Second Edition.
Stephen W. Freiman and John J. Mecholsky, Jr.

EARLY BRITTLE FRACTURE RESEARCH

Although fracture studies of ceramics can be traced as far as back as 1867, our current understanding of brittle fracture can be traced to Inglis (1913), who initiated the concept of stress concentration at a void in a material as shown in Figure 2.1.

Assuming an elliptically shaped cavity, the concentrated stress, σ_c, due to the presence of the void is given by the following expression:

$$\sigma_c = \sigma\left(1+2\sqrt{\frac{a}{\rho}}\right) \tag{2.1}$$

where $2a$ is the long axis of the ellipse and ρ is its radius of curvature. As ρ approaches zero, i.e. the elliptical cavity begins to resemble a crack whose tip radius is of atomic dimensions. The atomistic nature of materials and the nonlinearly elastic or plastic deformation that occurs in the vicinity of a crack tip avoid the problem of $\rho \rightarrow 0$, but this approach does not yet allow us to quantify the stress state at an actual crack tip and says nothing about when the void is likely to grow.

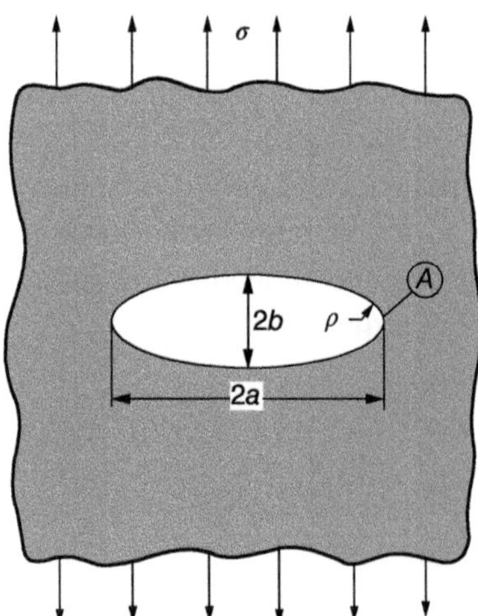

Figure 2.1 Void in a plate in which stresses are concentrated at point A.

The second important advance in our understanding of brittle failure was made by Griffith (1920) who postulated that brittle failure in glass is a result of the growth of small cracks when the material is subjected to a large enough tensile stress. He put forth the hypothesis that these cracks are present in all glasses in a distribution of sizes, leading to the concept that the smaller the volume (or area) under stress, the less likelihood of finding a large flaw and therefore the higher the strength. Griffith demonstrated this concept experimentally by measuring the strength of glass fibers of varying diameter and showing that strength increased with decreasing fiber diameter. The flaw that eventually grows to failure is determined by its severity as well as its location with respect to the highest tensile stress, thereby giving rise to the statistical nature of brittle failure.

Griffith also hypothesized that a material's resistance to the growth of a crack is determined by the energy required to create the two fracture surfaces produced by its extension. This approach assumes that fracture occurs in an equilibrium manner, i.e. in the absence of any kinetic effects, and that no energy is lost due to plastic flow or heat. It also neglects possible effects of the test environment in which the flaw is growing, e.g. water. Griffith's expression for glass fracture based upon this approach is given by

$$\sigma_f = \left(\frac{2E\gamma_f}{\pi a} \right)^{1/2} \tag{2.2}$$

where σ_f is the fracture strength, E is Young's modulus, γ_f is the energy required to form the crack surfaces (i.e. the fracture energy), and a is the critical flaw size. While Griffith's calculations were not entirely accurate because of his assumption that the fracture energy of the glass could be extrapolated from measurements of surface energy carried out at elevated temperatures, the form of Eq. (2.2) accurately depicts the relationship between strength, flaw size, and fracture energy. However, what was still required was a way of translating the external loads on a part into knowledge that could be used to predict its resistance to crack growth. This awaited the development of fracture mechanics.

DEVELOPMENT OF FRACTURE MECHANICS

Credit for the development of the science of fracture mechanics is rightfully given to George Irwin and his colleagues at the US Naval Research Laboratory (Irwin 1958). Irwin first introduced the concept of a "strain energy release rate," G, which has nothing to do with time dependence, but is the change in strain energy with crack extension. Equation (2.2) then becomes

$$\sigma_f = \left(\frac{EG}{\pi a}\right)^{1/2} \tag{2.3}$$

$$G = 2\gamma f \tag{2.4}$$

There are three ways that stress can be applied to a crack (Figure 2.2).

The predominant situation with respect to the stressing of brittle materials leads to crack growth under mode I loading, i.e. pure tension across the crack face.

A schematic of the stress state at the crack tip produced by an externally applied mode I load is shown in Figure 2.3.

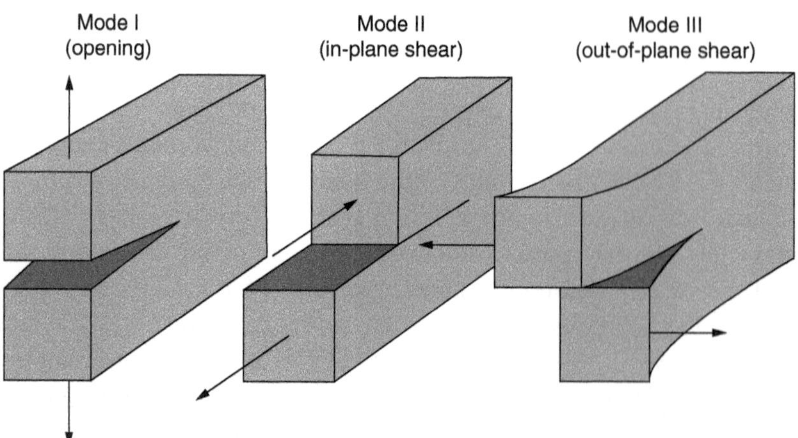

Figure 2.2 Modes of fracture. Source: From Anderson (1995). Reproduced with permission of Taylor & Francis.

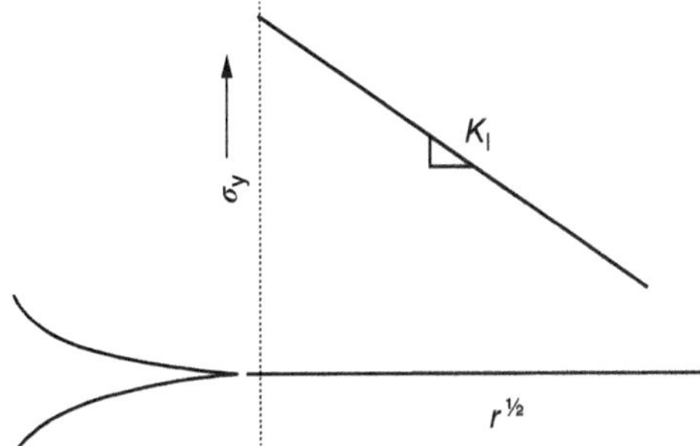

Figure 2.3 Schematic of the stress field at a crack tip.

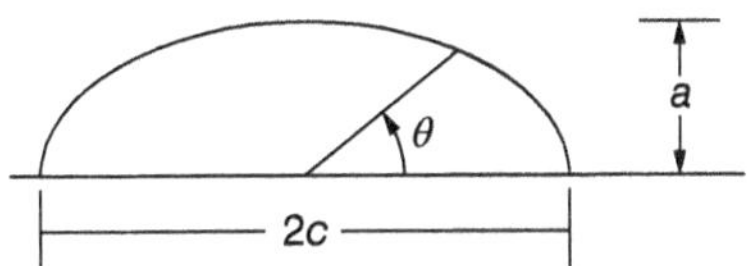

Figure 2.4 Surface flaw.

The stress intensity factor is defined through the following expression:

$$\sigma_y = \frac{K_I}{\sqrt{2\pi r}} f(\phi) \tag{2.5}$$

where σ_y is the stress at the crack tip, K_I is the slope of the stress-distance plot, ϕ is the angle in the plane with respect to the crack face, and r is the distance from the crack tip. Note that Eq. (2.5) is only valid outside of the singularity-dominated zone, within which the material is nonlinearly elastic or in which some permanent deformation has taken place. The beauty of linear elastic fracture mechanics is that it can be used to explain fracture in spite of the existence of this singularity.

The relationship between K_I and applied stress can take many forms depending on the type of loading and the location of the crack relative to the load. The simple example of a surface flaw, the predominant source of failure for brittle materials (Figure 2.4), gives rise to the relationship

$$K_I = Y\sigma a^{1/2} \tag{2.6}$$

where Y is a numerical factor that depends on the loading method and the crack geometry and a is the depth of the crack into the interior.

The surface flaw is typically elliptical rather than semicircular in shape and is usually elongated along the tensile surface. The expression for stress intensity factor for such a flaw is given by

$$K_{\mathrm{I}} = \sigma M\left(\frac{\pi a}{\Phi^2}\right)^{1/2} \tag{2.7}$$

where Φ is an elliptical integral that accounts for the fact that the stress intensity factor varies around the perimeter of the flaw and is given by

$$\Phi = \int_0^{\pi/2}\left[\sin^2\theta + \frac{a^2}{c^2}\cos^2\theta\right]^{1/2} \mathrm{d}\theta \tag{2.8}$$

The stress intensity is a maximum at the tip of the minor axis of the flaw and is a minimum at the tip of the major axis. M is a surface correction factor whose value is approximately 1.12 as determined by finite element analyses (Raju and Newman 1979).

CRITICAL FRACTURE TOUGHNESS

Crack extension takes place when the applied load rises to a level such that the stress intensity factor produces crack tip stresses large enough to rupture atomic bonds at the crack tip. *Fracture toughness is defined as the value of the stress intensity factor, K_{I}, at the point where $K_{\mathrm{I}} = K_{\mathrm{IC}}$, the critical stress intensity factor:*

$$K_{\mathrm{IC}} \equiv \left(EG_{\mathrm{C}}\right)^{1/2} \equiv \left(2E\gamma_{\mathrm{f}}\right)^{1/2} \tag{2.9}$$

Equation (2.9) provides the link between fracture mechanics and Griffith's fracture energy.

How do we measure criticality? Griffith defined γ_{f} in thermodynamic terms, but for most materials the point of reversible crack extension and contraction is not observed. Experimentally, we can increase the stress on a crack to the point at which the crack grows to failure. However, most brittle materials are sensitive to the effects of the atmosphere in which they are tested. Subcritical crack growth, otherwise known as stress corrosion, or environmentally enhanced crack growth is very difficult to avoid (see Chapter 3). The key is the relative humidity of water in the environment, i.e. its partial pressure. No liquid, no matter how small is the solubility of water in it, can be considered to be completely inert. Even in relatively neutral external environments, e.g. dry gases and vacuum, some materials will still undergo crack growth prior to catastrophic fracture due to other crack tip mechanisms. K_{IC} for brittle materials has at times been defined as the point at which rapid crack extension occurs, usually seen as a drop in the load on a testing machine. But then how rapid is rapid?

There are two options with respect to defining fracture toughness: (i) one can assume that K_{IC} is a fundamental material property and that each fracture mechanics technique represents an attempt to estimate its value by a different method, or (ii) one could assume that fracture toughness can *only* be defined in terms of the measurement itself and that there is nothing particularly fundamental about each determination. For the purposes of this document, we will assume the former, namely, that there is a number that we can call K_{IC} for each material, but whose determination will depend on many factors including test procedure, environment, etc.

Nonetheless, we are clearly dealing with an operational definition (as is also true for metals). For a given material, K_{IC} will be a function of environment, testing rate, and test geometry. For most of the tests described in this document, K_{IC} is calculated from the maximum load reached during the test. The crack velocity at which this point occurs will be a function of the compliance of both the specimen and the testing machine, but is usually in the range of 10^{-2}–10^{0} m s^{-1}. Fortunately, crack growth curves are typically very steep, so any uncertainty in crack velocity translates into only a small uncertainty in fracture toughness.

Another issue is the stress state at the crack tip, i.e. plane strain or plane stress. Whether there exists a state of plane stress or plane strain at the crack tip depends on the constraints. Where there is significant constraint, e.g. thick specimens, and little or no plastic deformation, a state of plane strain will exist. In all specimens and materials, the outer surface of the specimen will be in a state of plane stress, but the depth of the plane stress zone could be quite small. For metals, the condition of plane strain can be defined in terms of the size of the crack tip singularity, which is related to the ratio of K_I to the yield stress in the material. Given that yield stresses for brittle materials are so large (basically immeasurable), zone sizes will be vanishingly small. It can be assumed that for almost any brittle material and for any specimen size, the crack will be in a state of plane strain.

The issue of plane strain or plane stress arises in particular when one transforms measurements between a fracture energy and a stress intensity approach. It should be noted that both approaches are equivalent; neither is more fundamental. The relationship between fracture toughness and fracture energy is given by

$$\text{In plane stress}: K_I = \sqrt{GE} \tag{2.10}$$

$$\text{In plane strain}: K_I = \sqrt{GE\left(1-\nu^2\right)} \tag{2.11}$$

where ν is Poisson's ratio. However, historically the $(1-\nu^2)$ term has been ignored in the literature in discussing brittle fracture. For most materials neglecting this term will introduce an ~4% uncertainty into the absolute value of either G or K_I depending on the primary measurement.

Given the above discussion, we suggest the following as a good working definition of fracture toughness.

For a brittle material in which no subcritical crack growth occurs, K_{IC} is the stress intensity factor at which crack growth begins.

MIXED-MODE LOADING

However, "mixed-mode" fracture can also occur if the face of the flaw is oriented at an angle to the tensile stress. A schematic of such a situation is shown in Figure 2.5. Both K_{II} and K_{III} components of the stress field can act on the flaw, with K_{II} being greatest at the tensile surface and K_{III} being largest at the maximum flaw depth.

Two issues arise in mixed-mode loading. The most important is the choice of a model that accurately combines the modes; the other issue is the possible existence of slow crack growth. Effects of mode III loading generally are not a significant factor in brittle fracture and usually can be ignored.

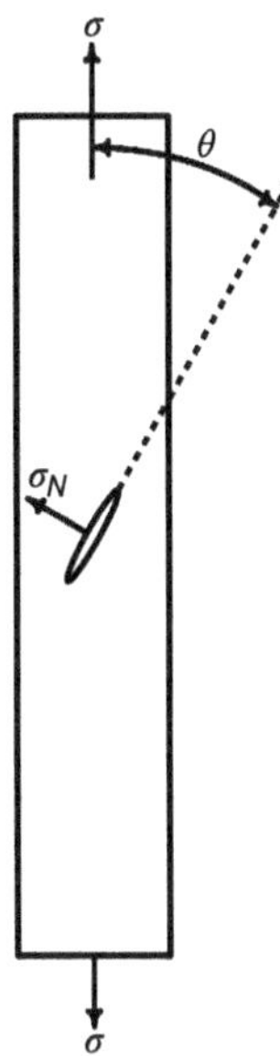

Figure 2.5 Schematic of mixed-mode loading.

Petrovic and Mendiratta (1976, 1977) discussed their results on the basis of three expressions for combined mode I and mode II stress intensity factors, the maximum normal stress criteria (Kordisch et al. 1976), the coplanar maximum strain energy release rate criteria (Erdogan and Sih 1963), and the noncoplanar maximum strain energy release rate criteria (Paris and Sih 1965) (Eqs. (2.12)–(2.14), respectively):

$$K_c = \cos\left(\frac{\theta}{2}\right)\left[K_I \cos^2\left(\frac{\theta}{2}\right) - \frac{3}{2}K_{II}\sin\theta\right] \tag{2.12}$$

$$K_c = \left(K_I^2 + K_{II}^2\right)^{1/2} \tag{2.13}$$

$$K_c = \left(\frac{2}{3+\cos^2\theta}\right)\left(\frac{1-\theta/\pi}{1+\theta/\pi}\right)^{\gamma/2\pi} \left[\left(1+3\cos^2\theta\right)K_I^2 + 8\sin\theta\cos\theta K_I K_{II} + \left(9-5\cos^2\theta\right)K_{II}^2\right]^{1/2} \tag{2.14}$$

where θ is the angle between the initial crack plane and the plane in which the crack propagates. Sih (1974) analyzed mixed-mode fracture and concluded that a crack will propagate in the direction of minimum strain energy density, S_{cr}, which can be expressed as

$$\begin{aligned} S_{cr} &= \frac{1}{\pi}\left(a_{11}K_{\mathrm{I}}^2 2a_{12}K_{\mathrm{I}}K_{\mathrm{II}} + a_{22}K_{\mathrm{II}}^2\right) \\ &= \frac{(\eta-1)K_{\mathrm{IC}}^2}{8\pi G} \end{aligned} \tag{2.15}$$

so that

$$K_{\mathrm{IC}}^2 = \frac{8G}{\eta-1}\left(a_{11}K_{\mathrm{I}}^2 + 2a_{12}K_{\mathrm{I}}K_{\mathrm{II}} + a_{22}K_{\mathrm{II}}^2\right) \tag{2.16}$$

where G is the shear modulus, $\eta = 3-4\nu$, for plane strain conditions and a_{11}, a_{12}, and a_{22} are given by

$$\begin{aligned} a_{11} &= \frac{1}{16G}\left[(1+\cos\theta)(\eta-\cos\theta)\right] \\ a_{12} &= \frac{\sin\theta}{16G}\left[2\cos\theta-(\eta-1)\right] \\ a_{22} &= \frac{1}{16G}\left[(\eta+1)(1-\cos\theta)+(1+\cos\theta)(3\cos\theta-1)\right] \end{aligned} \tag{2.17}$$

In the study by Petrovic and Mendiratta (1976, 1977), the flaws were nearly semicircular. In many cases, it is often necessary to use an expression to calculate K_{I} and K_{II} that takes into account the degree of ellipticity of the flaw. For K_{I} this expression is shown in Eq. (2.18) and is a relatively simple procedure, well known in the fracture mechanics literature (Randall 1967):

$$K_{\mathrm{I}} = \frac{Y}{\phi}\sigma c^{1/2}\sin^2\theta \tag{2.18}$$

where $Y = (1.2\pi)^{1/2}$ for a surface flaw, θ is the angle between the stress direction and the initial crack plane, ϕ is an elliptical integral of the second kind involving the major (b) to minor (a) radii ratio of the flaw, and c is the smaller of a or b. For K_{II}, an expression used by Petrovic and Mendiratta (1976, 1977) can be used:

$$K_{II} = \left[\frac{4}{\pi^{1/2}(2-\nu)}\right]\sigma c^{1/2}\sin\theta\cos\theta \tag{2.19}$$

Equations (2.18) and (2.19) yield values of K_I and K_{II} at the boundary of the flaw and the specimen surface. K_c can also be calculated by measuring fracture mirror size (Mecholsky et al. 1974, 1977):

$$K_c = \frac{K_{bj}}{\phi}\left(\frac{c}{r_j}\right)^{1/2} \tag{2.20}$$

where K_{bj} is the branching stress intensity constant (i.e. "fracture mirror" constant) for the material ($=Y\sigma rj^{1/2}$), corresponding to either the mirror–mist (j = 1) or mist–hackle (j = 2) mirror radius, r; c is $(ab)^{1/2}$; and ϕ is an elliptical integral of the second kind and is a function of the a/b ratio.

Marshall (1984) studied the behavior of indented cracks with and without residual stress. He found contrasting responses of as-indented and annealed indentations to mixed-mode loading. As indented flaws, which are influenced by residual contact stresses, extend stably during failure testing and align normal to the maximum tension prior to failure. The critical configuration does not involve mixed-mode loading, but the strength is influenced by the resultant kinking of the crack. Annealed indentations, which are free of residual stresses, extend unstably at a critical loading with an abrupt change of fracture plane from that of the initial flaw. These results would be expected to apply generally to brittle materials as long as crack growth is not disrupted by microstructural variations on a scale comparable with the crack size. He found that all of the fracture criteria that are consistent with observed crack paths provide underestimates of the strengths of stress-free indentation flaws in mixed-mode loading.

Marshall (1984) observed that the most soundly based fracture criterion (maximum strain energy release rate) underestimates the strength of the annealed flaws in mixed-mode loading, as does the maximum normal stress criterion. The strain energy density and coplanar strain energy release rate criteria also underestimate the strengths of the annealed indentation flaws, but both criteria give predictions that are consistent with the data for the annealed machining damage. Marshall attributed the strength deviation in the annealed specimens to possible shear tractions due to surface asperities. Glaesemann et al. (1987) suggested that mixed-mode cracks will reorient themselves to be perpendicular to the maximum principal tensile stress due to slow crack growth processes for both indented and stress-free cracks. In materials that exhibit slow crack growth, this would explain the reason for the strength to be less than that predicted by mixed-mode theory. Furthermore, Marshall (1984) determined an empirical correlation between the strengths of as-formed flaws in mixed-mode loading and predictions based on the coplanar strain energy release rate criterion:

$$\frac{\sigma_\theta}{\sigma_0} = (\cos\theta)^{-1/2} \tag{2.21}$$

Although there is no physical basis for this correlation (as-formed flaws are perpendicular to the applied stress at the point of failure), it is useful for making predictions of strength due to mixed-mode loading. Obviously more work is needed in the area of mixed-mode loading to translate from testing to the understanding of mechanisms.

Another, more recent and promising criterion for predicting the behavior of brittle elastic bodies under the influence of mixed-mode loading is known as the generalized maximum tangential stress (GMTS) criterion (Smith et al. 2001; Ayatollahi and Aliha 2009). The tangential stress, $\sigma_{\theta\theta}$, in a linear elastic cracked body can be written as an infinite series expansion (Williams 1957):

$$\sigma_{\theta\theta} = \left[\frac{1}{(2\pi r)^{1/2}}\right]\left(\cos\frac{\theta}{2}\right)\left(K_{\mathrm{I}}\cos\frac{\theta}{2} - K_{\mathrm{II}}\sin\theta\right) + T\sin^2\theta + \mathrm{O}(r^{1/2}) \tag{2.22}$$

where r and θ are the crack tip coordinates. T is usually called the T-stress and is a constant nonsingular term independent of the r distance from the crack tip. The $\mathrm{O}(r^{1/2})$ terms are higher order terms and are usually negligible near the crack tip. According to the GMTS criterion, crack growth initiates radially from the crack tip along the direction of maximum tangential stress, θ_0. Also the crack extension takes place when the tangential stress $\sigma_{\theta\theta}$ along θ_0 and at a critical distance r_c from the crack tip attains a critical value $\sigma_{\theta\theta c}$. Both r_c and $\sigma_{\theta\theta c}$ are assumed to be material constants.

The criterion for crack propagation under mixed-mode loading according to the GMTS theory is

$$K_{IC} = \cos\frac{\theta}{2}\left[K_{If}\cos^2\left(\frac{\theta_0}{2}\right) - \frac{3}{2}K_{IIf}\sin\theta_0\right] + \left(2\pi r_c\right)^{1/2} T_f \sin^2\left(\frac{\theta_0}{2}\right) \tag{2.23}$$

where K_{If}, K_{IIf}, and T_f are the critical values of the stress intensity and T-stress for mixed-mode fracture. The GMTS criterion has only been compared to a few materials. It should be used to test the results of previous researchers to see if this criterion can explain previous results.

The question of how the effects of the various modes should be combined mathematically to calculate fracture toughness is still a question of interest. Test procedures that can be used to determine crack growth characteristics under mixed-mode loading are described in Chapter 4.

SUMMARY

The strength of brittle materials is limited by the size and geometry of surface and volume defects. The most common defects are small surface flaws introduced during machining or handling. Linear elastic fracture mechanics provides a way to analyze and understand brittle fracture by linking applied stress to flaw severity through a stress intensity factor. We define critical fracture toughness as the stress intensity factor at which crack growth begins. Cracks can propagate under mixed modes of loading as well.

QUESTIONS

1. Compare and contrast Griffith's crack growth model and that developed by Irwin.
2. If a material fails with a strength of 14 MPa and a fracture energy of 8 J m^{-2} in tension, what is the size of the fracture initiating defect according to

Griffith's theory? According to Irwin's theory using the stress intensity approach? Assume $E = 450\,\text{GPa}$ and $Y = 1.26$.

3. Show that the singularity-dominated zone, r, is directly related to the crack size for an infinitely sharp crack in a brittle material.

REFERENCES

Anderson, T.L. (1995). *Fracture Mechanics, Fundamentals and Applications*, 2e. Boca Raton, FL: CRC Press.

Ayatollahi, M.R. and Aliha, M.R.M. (2009). Mixed mode fracture in soda lime glass analysed using the generalized MTS criterion. *Int. J. Solids Struct.* 46: 311–321.

Erdogan, S.F. and Sih, G.C. (1963). Crack extension in plates under plane loading and transverse shear. *J. Basic Eng.* 85: 519–527.

Glaesemann, G.S., Ritter, J.E. Jr., and Karl, J. (1987). Mixed mode fracture in soda lime glass using indentation flaws. *J. Am. Ceram. Soc.* 70 (9): 630–636.

Griffith, A.A. (1920). The phenomena of rupture and flow in solids. *Philos. Trans. R. Soc. Lond.* A221: 163–198.

Inglis, C.E. (1913). Stresses in a plate due to the presence of cracks and sharp corners. *Trans. Inst. Nav. Arch.* 55: 219–230.

Irwin, G.R. (1958). Fracture. In: *Handbuch der Physik*, vol. 6 (ed. S. Flugge), 551–590. Berlin–Heidelberg: Springer-Verlag.

Kordisch, H., Seidelmann, U., Soltedz, U., and Sommer, E. (1976). Studies of fracture behavior of combined normal and fracture-resistant ceramics. *Dtsh. Ver. Mater.* 6–7: 43–47.

Marshall, D.B. (1984). Mechanisms of failure from surface flaws in mixed-mode loading. *J. Am. Ceram. Soc.* 67: 110–116.

Mecholsky, J.J. Jr., Rice, R.W., and Freiman, S.W. (1974). Predictions of fracture energy and flaw size in glasses from mirror size measurements. *J. Am. Ceram. Soc.* 57: 440–444.

Mecholsky, J.J. Jr., Freiman, S.W., and Rice, R.W. (1977). Effect of grinding on flaw geometry and fracture of glass. *J. Am. Ceram. Soc.* 60: 114–117.

Munz, D. and Fett, T. (1999). *Ceramics*. Berlin–Heidelberg: Springer-Verlag.

Paris, P.C. and Sih, G.C. (1965). Stress analysis of cracks. In: *Fracture Toughness Testing and Its Applications*, 30–81. Philadelphia, PA: American Society for Testing and Materials. Spec. Tech. Publ. No. 381.

Petrovic, J.J. and Mendiratta, M.G. (1976). Mixed-mode fracture from controlled surface flaws in hot-pressed Si,N. *J. Am. Ceram. Soc.* 59: 163–167.

Petrovic, J.J. and Mendiratta, M.G. (1977). Correction of mixed-mode fracture from controlled surface flaws in hot-pressed Si_3N_4. *J. Am. Ceram. Soc.* 60: 463.

Raju, I.S. and Newman, J.C. Jr. (1979). Stress-intensity factors for a wide range of semi-elliptical surface cracks in finite-thickness plates. *Eng. Fract. Mech.* 11: 817–829.

Randall, P.N. (1967). In plane strain crack toughness testing of high-strength metallic materials. *ASTM STP* 410: 88–126.

Sih, G.C. (1974). Strain-energy density factor applied to mixed mode crack problems. *Int. J. Fract.* 10: 305–331.

Smith, D.J., Ayatollahi, M.R., and Pavier, M.J. (2001). The role of T-stress in brittle fracture for linear elastic materials under mixed mode loading. *Fatigue Fract. Eng. Mater. Struct.* 24: 137–150.

Williams, M.L. (1957). On the stress distribution at the base of a stationary crack. *J. Appl. Mech.* 24: 109–114.

CHAPTER 3

Environmentally Enhanced Crack Growth

INTRODUCTION

Because of its prevalence and importance in the design and selection of ceramic materials, we must introduce the phenomenon of environmentally enhanced crack growth in ceramics before discussing any measurement issues. Almost all glasses and ceramics held under a constant tensile stress in water-containing environments can fail after a period of time. The time to failure is inversely proportional to the applied stress (Figure 3.1). This phenomenon has been termed "delayed failure," "static fatigue," or "stress corrosion." The behavior of glasses and ceramics is in many ways similar to stress corrosion in metals, but the mechanisms can be quite different.

In this chapter we review environmentally enhanced crack growth behavior in glass and ceramic materials and briefly describe mechanisms by which this process occurs.

The Fracture of Brittle Materials: Testing and Analysis, Second Edition.
Stephen W. Freiman and John J. Mecholsky, Jr.

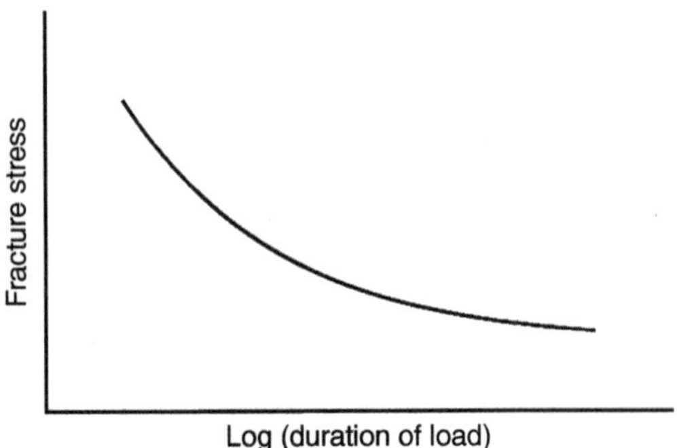

Figure 3.1 Schematic of delayed failure.

HISTORICAL BACKGROUND

The first observation of delayed failure was made in 1899 by Grenet (Grenet 1899), who loaded glass plates in three-point flexure, applying the stress by hanging a bucket from their centerline and gradually filling the bucket with water at varying rates (Figure 3.2). He observed that by decreasing the loading rate from $0.25\,kg\,min^{-1}$ to $0.25\,kg\,h^{-1}$ a 50%, decrease in breaking stress was obtained. Grenet also observed that specimens left under a constant load failed in times as long as five days.

There were many studies of this phenomenon over the ensuing 60+ years (Freiman et al. 2009). However, up until the development of fracture mechanics, the only available method of investigating delayed failure was measuring the strength of a specimen in flexure as a function of time under a constant load as Figure 3.1 depicts. These early investigators recognized that surface damage was important to the strength and that water and water vapor were an important factor in the failure process. However, because the small flaws in the surface of a specimen could not be directly observed, no details on the mechanism by which they were growing could be established.

Wiederhorn's (1967) paper on the effects of water on crack growth in soda-lime-silica glass paved the way for understanding the crack growth process. Wiederhorn used a double cantilever beam specimen, the details of which will be discussed in Chapter 4. The specimen was deadweight loaded, and crack extension was monitored through an optical microscope. By measuring the length of the crack as a function

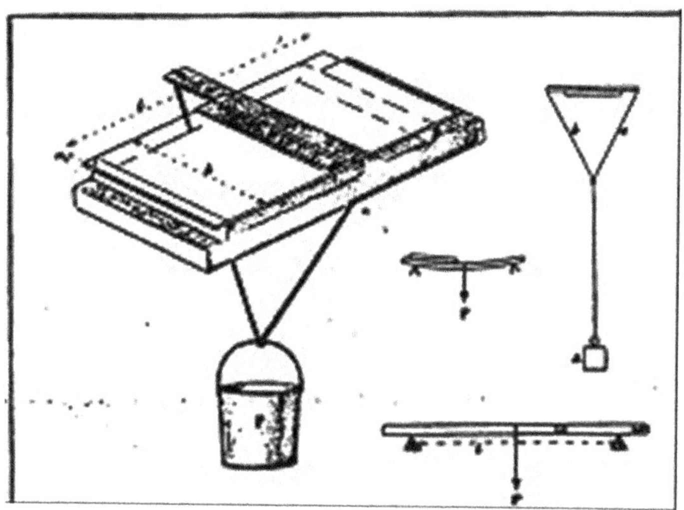

Figure 3.2 Schematic of experimental setup used by Grenet to first investigate delayed failure in glass.

of time, coupled with knowledge of the stress intensity factor at each crack length, a plot demonstrating the effects of various experimental parameters on crack growth behavior could be obtained.

MECHANISMS OF ENVIRONMENTALLY ENHANCED CRACK GROWTH

Figure 3.3 shows crack velocity as a function of stress intensity factor for soda-lime glass in N_2 gas of different water contents (relative humidities) (Wiederhorn 1967).

The curves in Figure 3.3 can be divided into separate regimes of behavior.

Region I

Region I is the regime of primary importance in terms of both chemical effects and for engineering design. Surface cracks in ceramic bodies spend most of their growth period in this portion of the curve. In this regime, crack growth rates are governed by the rate of reaction of an

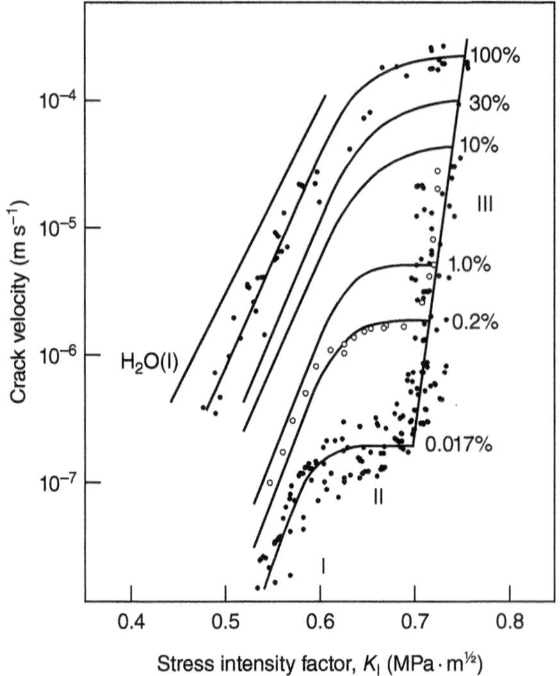

Figure 3.3 Crack growth data for soda-lime-silica glass. Source: After Wiederhorn (1967). Reproduced with permission of John Wiley & Sons.

environmental molecule, e.g. H_2O, with a highly strained crack tip bond. Wiederhorn (1967) demonstrated that chemical reaction rate theory can be used to obtain a rate equation for crack growth. By assuming that crack velocity is directly proportional to reaction rate, the following expression is obtained:

$$V = V_0 a(\mathrm{H_2O}) \exp\left(\frac{-E^* + bK_\mathrm{I}}{RT}\right) \quad (3.1)$$

where $a(\mathrm{H_2O})$ is the activity of water, all of the non-stress-related terms are included in the activation energy, E^*, and b is proportional to the activation volume for the chemical reaction, i.e. the increase in volume of a molecule between its unstressed state and that of the activated complex. At constant temperature, the slope of the V–K_I curve is directly proportional to the activation volume.

Because it is the chemical activity of the water, i.e. its partial pressure relative to saturation in a given environment, rather than its absolute concentration, completely inert environments are almost impossible to achieve. No liquid can be considered as completely inert even though the absolute quantity of water dissolved in it may be quite small.

Another variable affecting crack growth is the pH of aqueous solutions. As an example in Figure 3.4, one sees that the slope of the crack growth curve is much smaller when a soda-lime-silica glass is tested in basic solutions as opposed to acids. This behavior indicates that the OH^- ion is the critical component in the reaction.

The chemical composition of glasses and ceramics also has a marked effect on the position and slope of the crack growth curve (Figure 3.5). However, we have no way of systematically predicting either the location or the slope of such curves as materials are varied.

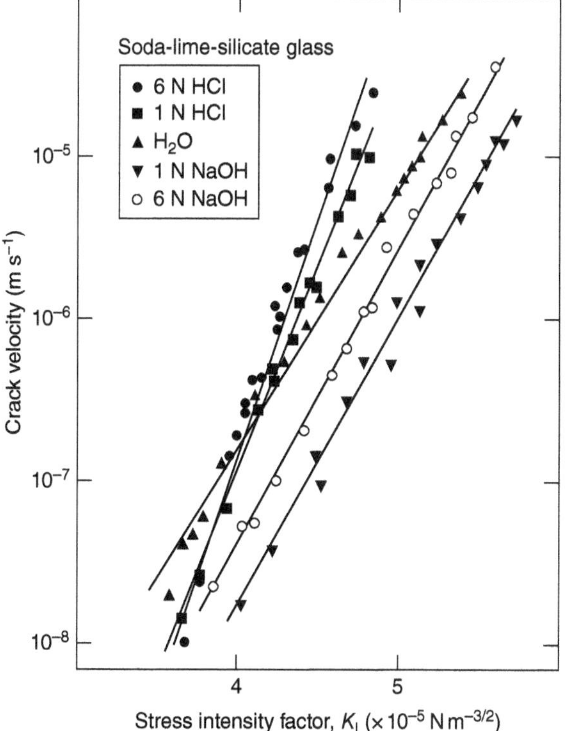

Figure 3.4 Effect of solution pH on crack growth curves for soda-lime-silica glass. Source: Wiederhorn and Bolz (1970). Reproduced with permission of John Wiley & Sons.

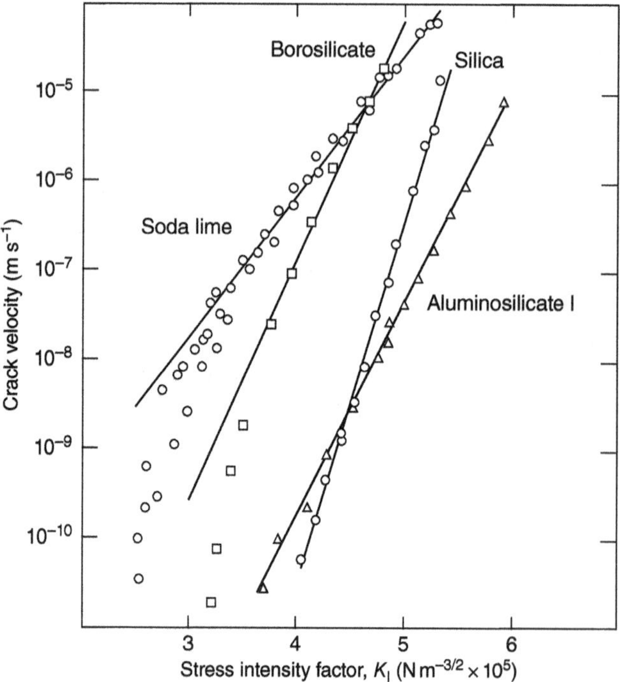

Figure 3.5 Effect of glass composition on crack growth behavior. Source: Wiederhorn and Johnson (1973). Reproduced with permission of John Wiley & Sons.

Region II

Region II is seen as more or less as a plateau in the V–K_I curve (Figure 3.3). In this regime, the crack is growing faster than the reacting molecule can reach the strained bonds at the crack tip. The rate-governing step now becomes the rate of diffusion of the active species through a water-depleted zone to the crack tip. Crack growth rates in this regime can be modeled using a Stokes–Einstein model (Wiederhorn et al. 1982):

$$V = \frac{0.0275kT}{(6\pi r\delta)(X_0/\eta)} \tag{3.2}$$

where r is the radius of water molecule, δ is the thickness of the boundary layer, and X_0/η is the ratio of concentration of water to solution viscosity. Notice that crack velocity is now directly proportional to the concentration (rather than activity) of the active species and inversely proportional to the viscosity of the solution.

Region III

Region III is the very steep portion of the V–K_I curve. While crack growth in this region is now independent of water content, there is still an effect of environment. Crack growth rates are governed by electrostatic interactions between the crack tip atoms and the surrounding environment (Wiederhorn et al. 1982). The slope of the crack growth curve in region III is dependent on the dielectric constant of the environment in which it is measured.

Crack Growth Thresholds

Some glasses form straight lines on a plot of the logarithm of the velocity versus the applied stress intensity factor over the measurable range of behavior, whereas others appear to exhibit a steep downturn at small K_Is, suggesting crack growth threshold, i.e. a stress intensity factor below which crack growth apparently ceases (Figure 3.5). Glasses containing mobile cations exhibit thresholds, whereas glasses containing no mobile ions, e.g. vitreous silica, do not.

Both Michalske (1977) and Gehrke et al. (1991) gave definitive evidence of a crack growth limit in glasses containing mobile ions by showing that a crack loaded at this K_I required additional time to repropagate when the load was increased. Depending on the time held at load, the time to repropagate was shown to be many hours or even days. The delay to failure decreased rapidly when the crack was loaded much higher or much lower than the crack growth limit. These results may be taken as evidence that the crack had really stopped propagating.

Depending on the glass composition, low-level plateaus in which crack growth seems to be independent of the applied stress intensity factor (Figure 3.6) have also been observed (Simmons and Freiman 1981). This behavior is attributed to the stresses at the crack tip generated by ion exchange.

For most materials cycling the load does not affect the rate of crack growth. It is only the tensile component of the stress that causes the

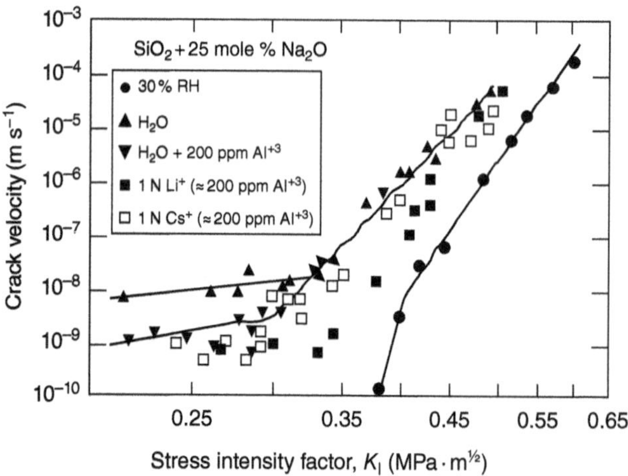

Figure 3.6 Low-level plateaus observed in binary Na_2O–SiO_2 glasses. Source: Simmons and Freiman (1981). Reproduced with permission of John Wiley & Sons.

environmentally enhanced process. However, for materials in which the microstructure can contribute to wedging-induced tensile stresses at the crack tip, crack growth under compression will take place (Evans 1980; Freiman et al. 2009).

Molecular Mechanism for Crack Growth in Oxide Glasses

Why is water particularly effective in promoting crack growth, and will other molecular species enhance the growth of cracks? Michalske and Freiman (1983) published a paper describing a molecular mechanism by which water and some nonaqueous environments can react with a strained Si—O bond. Their model is based on the schematic shown in Figure 3.7, which depicts a strained crack tip bond in a silicate glass. Of particular importance is the fact that a water molecule contains one set of atomic orbitals that do not participate in the oxygen–hydrogen bonding and can donate a proton as well. The three steps in the bond rupture process are:

1. A water molecule attaches to a bridging Si–O–Si bond at the crack tip. The water molecule is aligned by hydrogen bonding with the $O_{(bridging)}$ and interaction of the lone-pair orbitals from $O_{(water)}$ with Si.

2. A reaction occurs in which both proton transfer to the $O_{(br)}$ and electron transfer from the $O_{(w)}$ to the Si take place. It is here that the original bridging bond between $O_{(br)}$ and Si is destroyed.
3. Rupture of the hydrogen bond between $O_{(w)}$ and transferred hydrogen occurs to yield Si–O–H groups on each fracture surface.

In this model there is no requirement for prior dissociation of the water molecule, nor must any reaction products be removed. Second, the model suggests that environments other than water should enhance crack growth if the species possesses structural and bonding features similar to water, i.e. proton donor sites at one end and lone-pair orbitals at the other, as well as being of comparable size with the Si—O bond.

Michalske and Freiman demonstrated that molecules such as ammonia, hydrazine, and formamide, which have the requisite structure

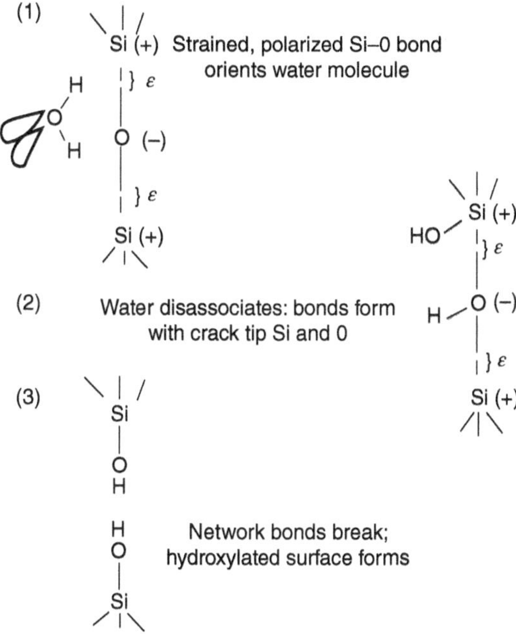

Figure 3.7 Schematic of stress-enhanced reaction between water and a strained crack tip bond in silica. Source: After Michalske and Freiman (1983). Reproduced with permission of John Wiley & Sons.

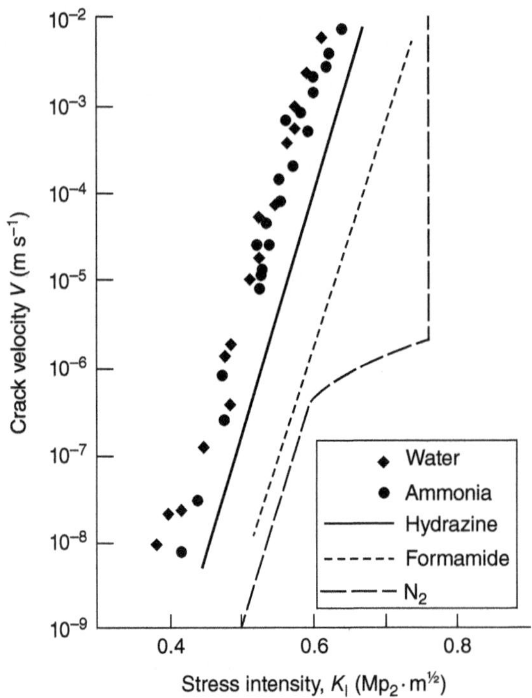

Figure 3.8 Crack velocity-K_I curves in environments that promote crack growth. Source: Michalske and Freiman (1983). Reproduced with permission of John Wiley & Sons.

noted above, were effective in enhancing crack growth rates (Figure 3.8). The absence of a region II plateau in these curves is the key to being able to conclude that it is the environment itself and not dissolved water that is causing the crack tip reaction.

Other molecules such as carbon monoxide and acetonitrile that do not have lone-pair orbitals opposite proton donor sites were not crack growth enhancing.

Modifier ions such as Na^+ and Ca^{++} increase the complexity of the crack growth plots (Figure 3.9) (Freiman et al. 1985) but do not change the basic mechanism of bond rupture.

Other Materials

Although it is known that almost all ceramics are subject to water-enhanced crack growth, little is known as to the possible effects of other

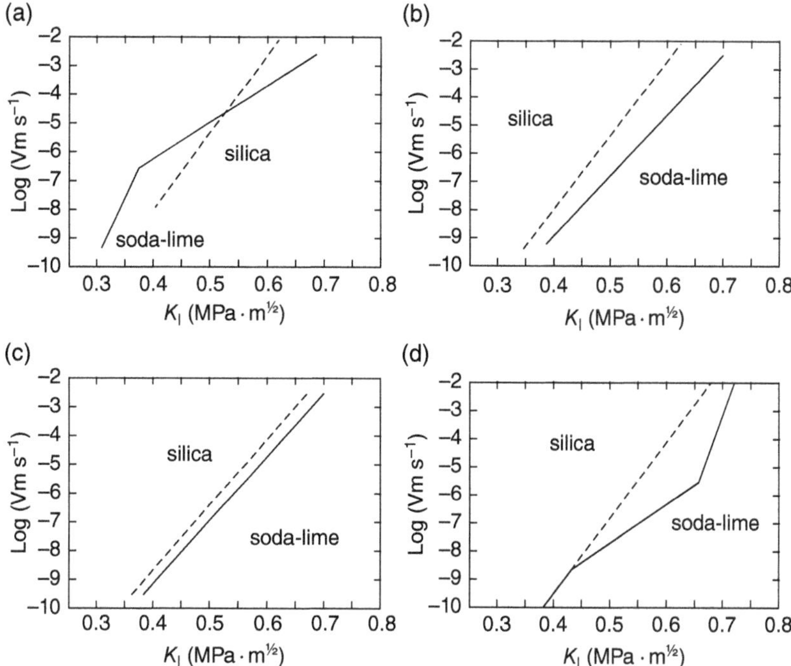

Figure 3.9 Crack growth in vitreous silica and soda-lime glass. Environments are (a) water, (b) ammonia, (c) hydrazine, and (d) formamide. Source: Freiman et al. (1985). Reproduced with permission of John Wiley & Sons.

environments. Only Al_2O_3 (sapphire single crystals) and MgF_2 (magnesium fluoride) have been fully investigated (Michalske et al. 1986). Except for the change in the slopes of the crack growth curves, sapphire follows the same behavior as silica (Figure 3.10).

Because it is an ionically bonded solid, MgF_2 behaves quite differently (Figure 3.11). Charge separation due to stressing is much less important. As a consequence, the presence of an actual liquid at the crack tip rather than simply an environmental molecule seems to be needed for cracks to grow. The dielectric constant of the liquid is the critical factor. Figure 3.12 shows a schematic of this type of behavior.

Finally, it should be noted that purely covalent materials such as silicon do not undergo any environmentally enhanced crack growth. In such materials stress does not change the degree of charge separation between Si atoms.

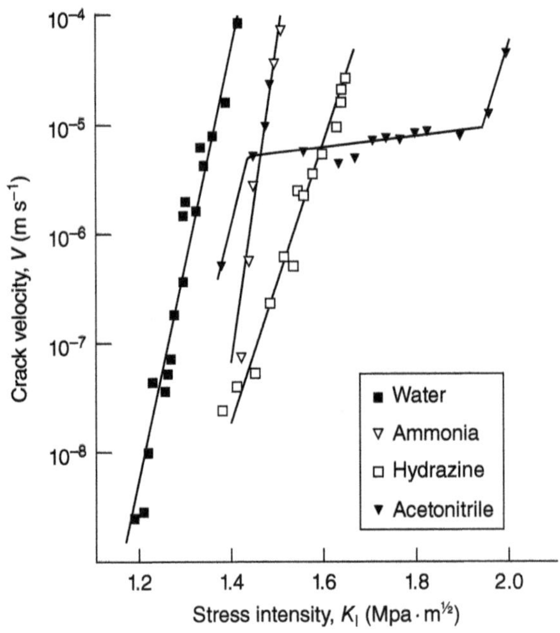

Figure 3.10 Environmental effects on crack growth in Al_2O_3. Source: Michalske et al. (1986). Reproduced with permission of John Wiley & Sons.

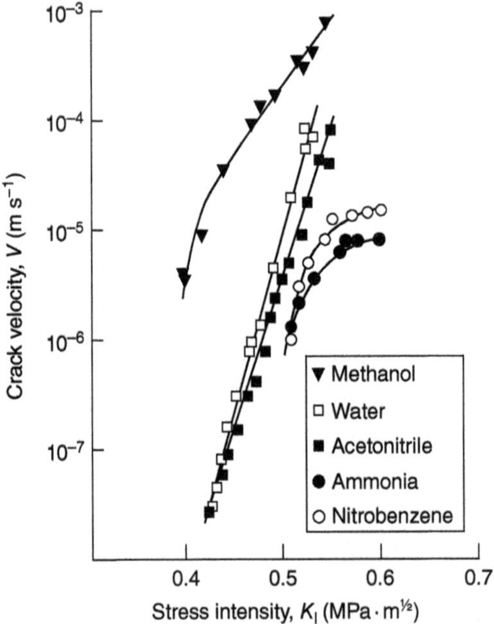

Figure 3.11 Environmental effects on crack growth in MgF_2. Source: Michalske et al. (1986). Reproduced with permission of John Wiley & Sons.

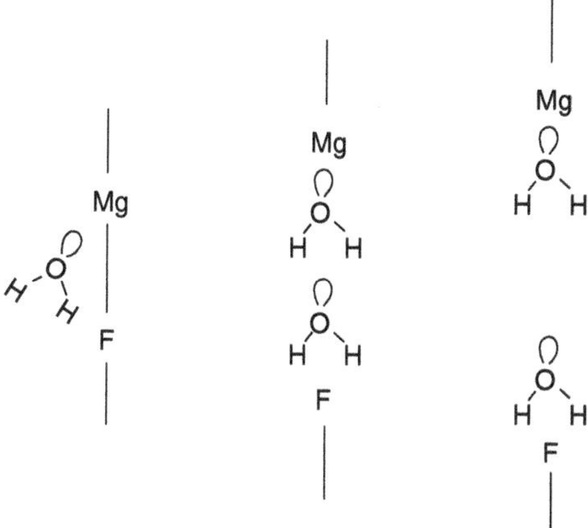

Figure 3.12 Proposed mechanism of water reaction in MgF_2. Source: Michalske et al. (1986). Reproduced with permission of John Wiley & Sons.

SUMMARY

Environmentally enhanced crack growth is a pervasive phenomenon in ceramics and related materials. For most materials the mechanism is a reaction between an external molecule and the strained bonds at a crack tip. Unfortunately, despite extensive experimental data on a variety of materials, there is no model that allows for the a priori prediction of the severity of crack growth for a given material.

QUESTIONS

1. Michalske and Freiman developed a theory for stressed silica glass in water that explained the mechanism of stress corrosion as a concerted acid–base reaction at the crack tip. Following their concepts, describe, as well as you can, the expected stress corrosion process for arsenic sulfide glass, As_2S_3. Recall that S is a chalcogenide, which means "oxygen-like." Use either words or schematic diagrams (preferred) with a few words to describe the process.
2. The fatigue limit in glasses is a subject of much discussion. Please provide a reasonable explanation of the behavior of soda-lime-silica glasses, silica glass, and soda-silicate glasses at low loads and low crack velocities in an active environment. In other words, contrast the behavior of these glasses in an aqueous environment.

3. Which liquids would be expected to cause stress corrosion in soda-lime-silica glass, and which liquids would not? Why?
4. Observation of stress corrosion in glasses shows that there are three regions of crack growth above the threshold limit. Briefly explain the reasons for these three regions in terms of the environment relative to the crack velocity.

REFERENCES

Evans, A.G. (1980). Fatigue in ceramics. *Int. J. Fract.* 16: 485–498.

Freiman, S.W., White, G.S., and Fuller, E.R. Jr. (1985). Environmentally enhanced crack growth in soda-lime glass. *J. Am. Ceram. Soc.* 69: 38–44.

Freiman, S.W., Wiederhorn, S.M., and Mecholsky, J.J. Jr. (2009). Environmentally enhanced fracture of glass: a historical perspective. *J. Am. Ceram. Soc.* 92: 1371–1382.

Gehrke, E., Ullner, C., and Hahnert, M. (1991). Fatigue limit and crack arrest in alkali-containing silicate glasses. *J. Mater. Sci.* 26: 5445–5455.

Grenet, L. (1899). *Bull. Soc. Encour. Ind. Nat./Ser.5/* 4: 838. For English translation see Preston, F.W. (1934). The time factor in the testing of glassware. *Glass Ind* 15: 33–43.

Michalske, T.A. (1977). The stress corrosion limit: its measurement and implications. In: *Fracture Mechanics of Ceramics*, vol. 5 (ed. R.C. Bradt, D.P.H. Hasselman and F.F. Lange), 277–289. New York: Plenum Press.

Michalske, T.A. and Freiman, S.W. (1983). A molecular mechanism for stress corrosion in vitreous silica. *J. Am. Ceram. Soc.* 66: 284–288.

Michalske, T.A., Bunker, B.C., and Freiman, S.W. (1986). Stress corrosion of ionic and mixed ionic/covalent solids. *J. Am. Ceram. Soc.* 69: 721–724.

Simmons, C.J. and Freiman, S.W. (1981). Effect of corrosion processes on subcritical crack growth in glass. *J. Am. Ceram. Soc.* 64: 683–686.

Wiederhorn, S.M. (1967). Influence of water vapor on crack growth in soda-lime glass. *J. Am. Ceram. Soc.* 50: 407–414.

Wiederhorn, S.M. and Bolz, L.H. (1970). Stress corrosion and static fatigue of glass. *J. Am. Ceram. Soc.* 53: 543–548.

Wiederhorn, S.M. and Johnson, H. (1973). Effect of electrolyte pH on crack propagation in glass. *J. Am. Ceram. Soc.* 56: 192–197.

Wiederhorn, S.M., Freiman, S.W., Fuller, E.R. Jr., and Simmons, C.J. (1982). Effect of water and other dielectrics on crack growth. *J. Mater. Sci.* 17: 3460–3478.

CHAPTER 4

Fracture Mechanics Tests

INTRODUCTION

Until the development of fracture mechanics, it was not possible to directly measure the resistance of a material to the growth of defects. There are now a number of accepted procedures for determining fracture toughness as well as measuring the growth rate of cracks as a function of the stress intensity factor. In this chapter we will discuss in some detail the various techniques available to both determine fracture toughness and investigate the mechanisms of crack growth. The chapter is divided into a series of classes of fracture mechanics specimens. Each class has its pluses and minuses. At its end, we encounter the issue mentioned in Chapter 2, namely, K_{IC} is a true materials' parameter or one that varies with the test procedure. As we will see, standards development organizations have taken the latter position.

The Fracture of Brittle Materials: Testing and Analysis, Second Edition.
Stephen W. Freiman and John J. Mecholsky, Jr.

DOUBLE CANTILEVER BEAM SPECIMENS

Background

Double cantilever beam (DCB) specimens were the first to be used to directly determine the crack growth resistance of brittle materials, both ceramics and polymers. Figure 4.1 is a schematic of the general DCB specimen geometry, so-called because each half of a specimen is analyzed as a cantilever on a fixed foundation.

Beam theory is used to relate the fracture energy, γ_A, to the bending force, F, applied to the end of the specimen. The energy stored in the specimen is equated to that required to form two new surfaces, leading to the expression

$$\gamma_A = \frac{F^2 L^2}{2EIt} \tag{4.1}$$

where L is the crack length; E is Young's modulus, which in single crystals is the value along the longitudinal axis of the specimen; t is the thickness of the specimen at the site of the crack; and I is the moment of inertia of one-half of the specimen about its longitudinal axis. I is given by

$$I = \frac{Wh^3}{12} \tag{4.2}$$

where W is the total specimen thickness and h is the half width.

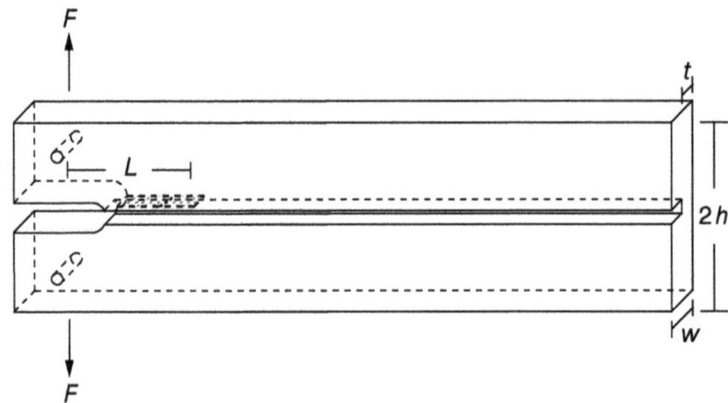

Figure 4.1 DCB specimen.

For specimens other than single crystals having well-defined cleavage planes, a guiding groove, ~1/2 of the specimen thickness, must be inserted along the centerline of the specimen to avoid the crack from breaking off one arm. Flat-bottomed grooves with sharp corners guide the crack most efficiently. The presence of a groove creates a number of issues. First, it changes the moment of inertia, I, of the specimen arms, which can lead to a miscalculation of the K_I. A simple way to account for the groove is to use the average dimension of each half of a specimen to calculate I for that half and average the two values of I. This method does not significantly reduce the accuracy of the K_I calculation as long as the ratio of groove width to specimen width is small. Wu et al. (1984) provide a more detailed correction of the effect of the groove on the moment of inertia.

Another issue created by grooving is the damage and possible residual stresses created during the cutting process. Heat treatments can sometimes be used to relieve machining-induced stresses in the groove.

Later work accounted for shear forces as well as the contribution to the strain energy in the uncracked portion of the specimen to yield a more accurate value of γ or K_{IC} (Eq. 4.3) (Wiederhorn et al. 1968):

$$\gamma = \gamma_A \left[1 + 1.32\frac{h}{a} + 0.54\left(\frac{h}{a}\right)^2 \right] \tag{4.3}$$

The DCB specimen shown in Figure 4.1 is loaded through a tensile force applied to pins or hooks inserted through holes drilled in the specimen. Loads can be applied by hanging weights on the loading pins through wires or by attaching the wires to a universal testing machine.

In this DCB configuration, the initial crack length is important to the accurate determination of fracture toughness. Measurements made for crack lengths $>1.5 \times h$ yield accurate results, but the crack must be no longer than $L = 1.5 \times h$ in order to avoid end effects. If h is too small, the crack will not grow down the center of the specimen but will turn to break off an arm; if h is too large, the forces required to propagate the crack can damage the specimen. Also, one must be able to accurately determine the position of the crack tip. Usually this is carried out by using a high intensity light that will reflect off of the crack face.

In opaque ceramics, however, this can be a significant problem. Even in translucent materials, e.g. alumina, under transmitted light, the crack tip can have a wavy appearance.

Applied Moment Loading

Suppose, instead of applying a force, we load a DCB specimen of the same general proportions as that in Figure 4.1 by applying a bending moment to the top of the specimen as shown in Figure 4.2. One can then show that G or K_I is crack-length independent by using Eq. (4.4) (Freiman et al. 1973), giving rise to the applied moment DCB (AMDCB) specimen

$$G = \frac{M^2}{EIt} \tag{4.4}$$

where M = (applied load, F)×(moment arm, l). The moment arm is fixed by the distance between the loading point and the fulcrum on

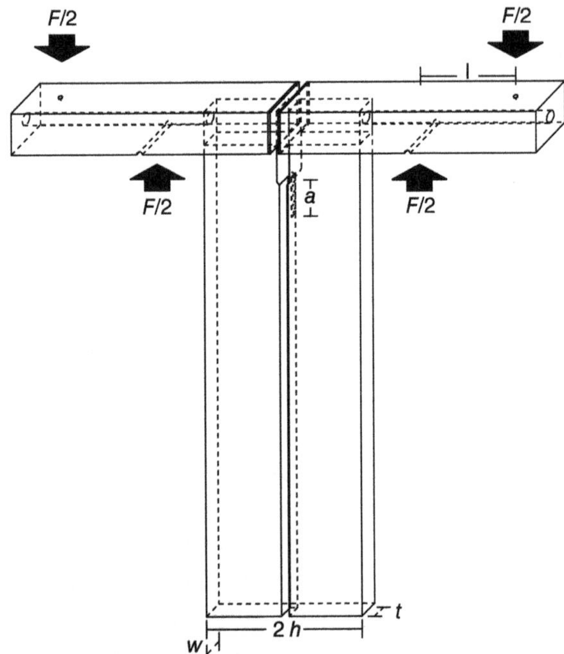

Figure 4.2 Applied moment DCB specimen.

the arms attached to the specimen. G can be converted to K_I through Eq. (2.6) to yield

$$K_I = \frac{M}{(It)^{1/2}} \tag{4.5}$$

Figure 4.3 shows the attachment of loading arms to the top of the specimen and a trapeze arrangement by which forces can be applied to the loading arms. The top portion of the fixture is hinged to allow

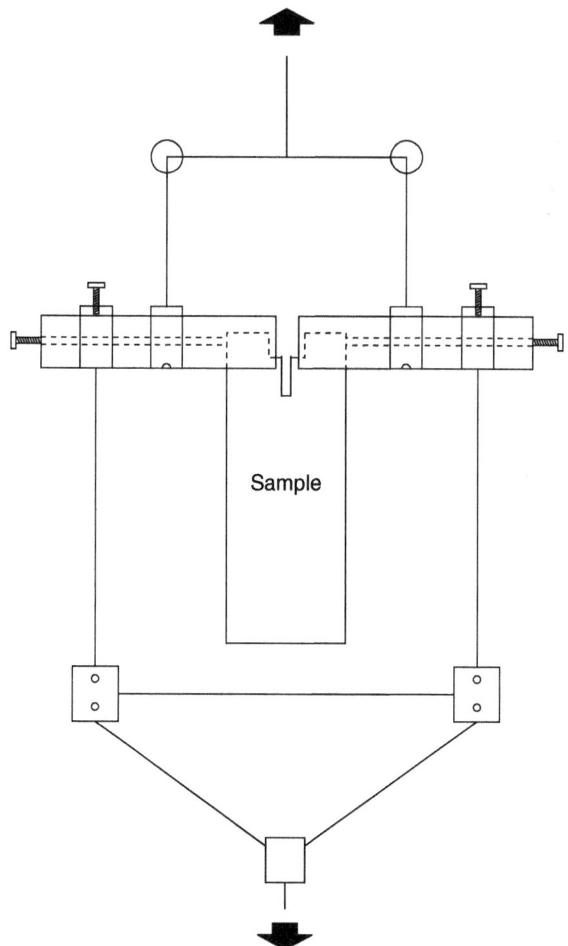

Figure 4.3 AMDCB specimen in loading fixture.

the arms to open as the crack propagates. The triangular portion at the bottom assures that the load is divided equally between the two sides. The trapeze is connected to the sample arms through wires attached to sleeves that slip over the arms; balls in the sleeves rest in dimples in the top of the arms.

Loads can be applied through a testing machine in the case of the determination of K_{IC} or through dead weights when crack velocity data is being obtained. In either case, care must be taken that the pins in the trapeze rotate with a minimum of friction because a small resistance to rotation can lead to significant overestimates of K_I.

Two types of loading arms are typically employed. The simplest is two metal bars with an opening at one end in which the specimen fits, a groove on one side to fit the trapeze, and a dimple on the opposite side to accommodate the loading point. These arms are cemented to the top of the specimen and work well for materials with a small resistance to crack growth, e.g. glasses, in which the loads on the arms are therefore relatively small. A second, more complex, type of arm involves an arrangement in which the arms are held to the top of the specimen through a clamping arrangement.

Contoured DCB

The contoured (or tapered) DCB configuration is an alternative means of making K_I crack-length independent. The strain energy release rate or crack driving force can be generally expressed as

$$G = \frac{P^2}{2t}\frac{\partial C}{\partial a} \tag{4.6}$$

where P is the applied load, t is the crack width, a is the crack length, and C is the specimen compliance or stiffness. If the DCB specimen can be contoured as shown in Figure 4.4, such that $\partial C/\partial a$ is constant, then the crack driving force is independent of crack length (Mostovoy et al. 1967).

K_I is given by

$$K_I = 2P\left(\frac{m}{tw}\right)^{1/2} \tag{4.7}$$

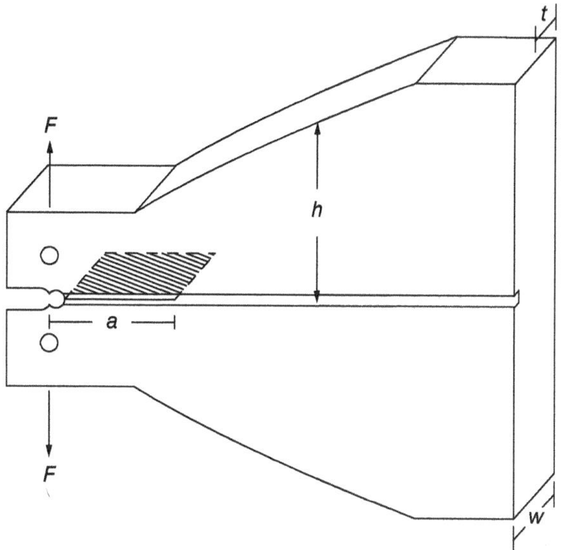

Figure 4.4 Contoured or tapered DCB specimen.

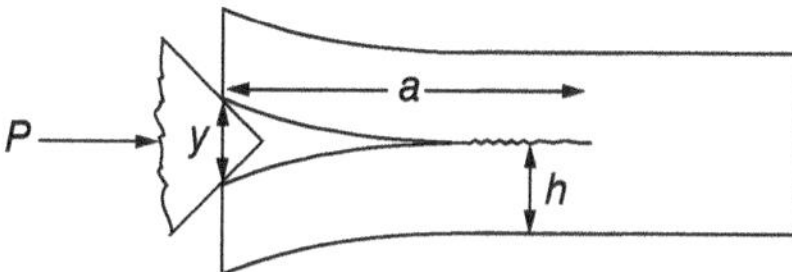

Figure 4.5 Wedge-loaded DCB.

$$m = \frac{3a^2}{h^3} + \frac{1}{h} \tag{4.8}$$

While the above expression is only totally accurate for a curved specimen, if *a/h* is large, then the approximation given by a linear taper yields relatively accurate values of K_L. Such specimens have been used in studying adhesion in joints and for polymeric materials but have not been employed extensively for inorganic materials for the most part because of the expense in machining these shapes.

Wedge-loaded DCB

DCB specimens can also be loaded by driving a wedge into the opening at the top of the specimen (Figure 4.5).

The crack grows stably into the specimen because it experiences a decreasing K_I field. In addition, because the specimen is loaded at the crack line, there is less tendency for the crack to deviate, reducing the need for grooving. The K_I expression for such a specimen is given by (Evans 1974)

$$K_I = \frac{\sqrt{3}}{2} \frac{Eyh^{3/2}}{a^2\left[1+0.64(h/a)\right]^2} \tag{4.9}$$

Uncertainty regarding frictional effects between the wedge and the specimen that affect the value of K_I means that this type of loading has not been used extensively for ceramics.

Introducing a Crack

In order to use any of the above configurations, one must first introduce a crack of the correct length into one end of the specimen. One method involves using a sharpened screw to put a force on the ungrooved side of the specimen just at the base of the starting notch. With care, crack pop-in occurs, but this procedure is sensitive to the applied force on the screw and results in the loss of a fair number of specimens. A better method is to use a Vickers or Knoop hardness indenter to place an indentation at the base of the notch and depend on the formation of the small cracks emanating from it to act as the starting crack in the specimen. The ungrooved side of the specimen should be polished to facilitate observation of a crack as it pops in. Using an intense light directed at an oblique angle to the specimen surface, one can see the crack as it initiates, even in opaque specimens of polycrystalline ceramics. Tapering the starting notch can also aid in the introduction of a stable crack.

Determination of K_{IC}

Let us assume that a specimen has been machined and a crack of the proper length introduced. How should one load this specimen and in what environment, in order to obtain a reproducible and precise, accurate measure of K_{IC}? As noted in Chapter 3, almost all ceramic materials are susceptible to moisture-induced slow crack growth. Because cracks grow at different rates depending on the moisture content of the atmosphere, measurements conducted slowly in air on such materials are likely to yield smaller values of K_{IC} than those taken in dry environments under rapid loading conditions. Because it is difficult, if not impossible, to predict such slow crack growth susceptibility for an untested material, the conservative approach is to conduct the test in relative dry environments, e.g. vacuum pump oil or dry nitrogen gas, and to load to failure as rapidly as the testing machine will permit. The maximum recorded load is used in the particular expression for K_{IC}. From a practical point of view, if the aim is to obtain data on the fracture resistance of a material to be used in air or some other environment, then measuring K_{IC} in that environment makes sense.

Measurement of Crack Growth Rates

The primary advantage of the DCB specimen is in conducting controlled crack growth experiments to obtain fundamental knowledge of fracture mechanisms. This means that the specimen is typically placed under a constant load and the position of the crack determined as a function of time. In most of the DCB configurations, one must account for the increase in K_I as the crack progresses, but in some, particularly the AMDCB, this is not required because K_I is independent of crack length.

The observation of a crack in transparent materials such as glasses and some single crystal and polycrystalline materials is fairly straightforward. A focused light is placed behind (or in front of) the specimen at an angle that will cause the crack face to be illuminated. A traveling microscope is then used to measure crack position. In glasses there is some scatter in the data from specimen to specimen due to variations of the angle that the crack makes with the specimen surface.

In translucent or opaque materials, crack growth measurements are more difficult, and greater scatter in data is observed. However, if the ungrooved side of translucent materials is polished, proper lighting can at times reveal the crack, but some accuracy in position is

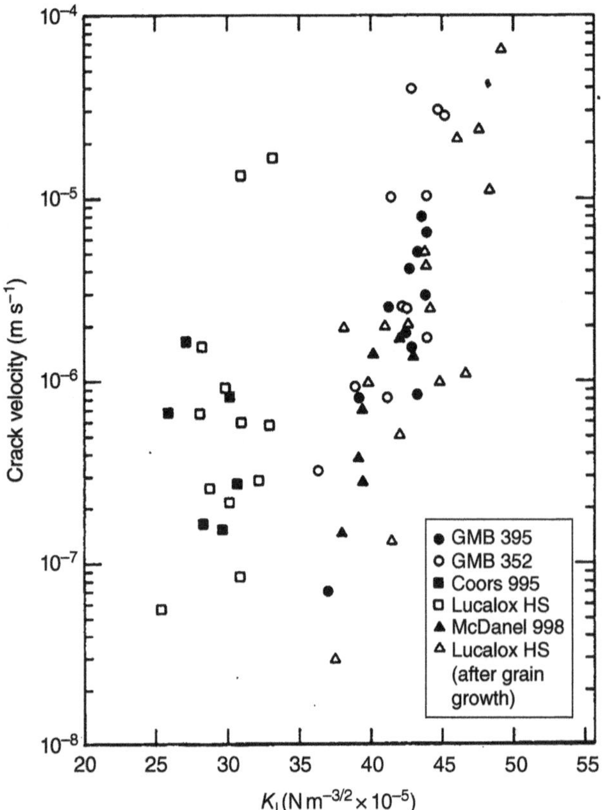

Figure 4.6 Crack growth data in various polycrystalline aluminum oxide materials taken using the AMDCB technique (Freiman et al. 1974).

lost. In completely opaque materials, the intersection of the crack with the polished face can appear as a line. As long as relative motion of the crack is all that is needed, as in the AMDCB specimen, data can be obtained, as shown in Figure 4.6 for several polycrystalline aluminas.

Some of the data scatter seen in Figure 4.6 results from the fact that the crack does not always grow smoothly through the thickness of the specimen. Because of the preference of a crack for a particular plane in a grain of the polycrystalline microstructure, one portion of the crack front may jump ahead then can slow down, while another piece of the crack grows.

DOUBLE TORSION TEST

Background

The double torsion (DT) test was first described by Outwater et al. (1974). Because the K_I in this configuration is also crack-length independent and all loading is carried out in compression, it was rapidly embraced as a test that would be particularly valuable for obtaining crack growth data in brittle materials, especially under conditions in which attachment of fixtures would be difficult. A review paper on the DT technique by Shyam and Lara-Curzio (2006) provides a wealth of detail on the background, analysis, and use of this technique.

As can be seen in Figure 4.7, the DT specimen is loaded by subjecting one end to a torque through the application of a four-point loading system. In the early versions of this test, a center groove was thought to be needed to prevent the crack from deviating from the center. Although the presence of such a groove can be included in the specimen analysis, it is not necessary; through proper alignment and loading, cracks will grow stably down the center of the specimen.

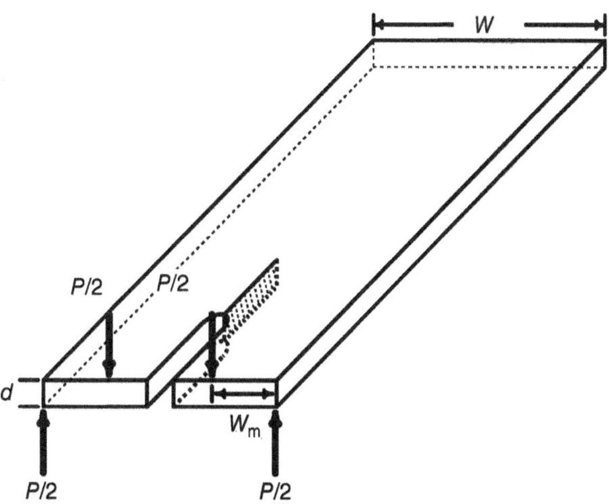

Figure 4.7 Schematic drawing of the double torsion specimen.

G and K_I

The driving force on the crack, i.e. the strain energy release rate, *G*, is given by Eq. (4.10):

$$G = \frac{3P^2 W_m^2 (1+\nu)}{Wd^3 d_w E} \tag{4.10}$$

where *P*, W_m, *W*, *d*, and d_w are defined in Figure 4.7. The Young's modulus is $E = 2G(1+\nu)$. Ψ is a factor determined by Fuller (1979) correcting for the fact that a finite beam is being analyzed and is given by

$$\Psi = 1 - 0.6302\tau + 1.20\tau \exp\left(-\frac{\pi}{\tau}\right) \tag{4.11}$$

where τ is a thinness ratio given by $(2d/W)$. The need for this correction comes about particularly in the use of thicker beams because of the potential interpenetration of the two sides of the specimen in torsion.

The stress intensity factor, K_I, for such a specimen is given by

$$K_I^2 = E'G \tag{4.12}$$

where $E' = E/(1-\nu^2)$ for plane strain. So that

$$K_I = PW_m \left[\frac{3(1+\upsilon)}{Wd^3 d_n \psi} \right]^{1/2} \tag{4.13}$$

K_I is independent of crack length over approximately the middle 60% of the specimen length. Too short a starting crack leads to an overestimate of the toughness, while too long a crack underestimates it.

The shape of the crack front depends on the crack growth characteristics of the material. There is continuing discussion as to whether the crack in a DT specimen is subject to pure mode I loading, but this does not appear to be a serious issue; values of fracture toughness obtained by the DT technique are comparable with those determined in other ways.

Precracking

As with the DCB technique, a starting crack must be introduced into the DT specimen before testing. The same crack introduction procedures applicable to the DCB test can be used. One frequently used method is to taper a notch at the front end of the specimen. If the notched specimen is loaded slowly, the pop-in of a crack can be observed through an observed drop in the load. A second method is to introduce a series of Knoop indentations along the starting notch.

Fracture Toughness Determination

Critical fracture toughness can be determined using the DT specimen by loading a precracked specimen at a rapid loading rate until the crack grows the length of the specimen. The initial crack tip must lie within the region of constant K_I. K_{IC} is calculated from Eq. (4.13) using the largest recorded load, P. The same issues of loading rate and moisture discussed with respect to the DCB test apply here as well.

While the DT method yields accurate values of fracture toughness, the quantity of material required for each specimen means that other techniques to determine K_{IC} may be better choices. The exception would be if the material is produced in large sheets or can be cast to shape such as with concrete. A more important use of the DT technique is in obtaining crack growth data in harsh environments or elevated temperatures where direct viewing of a crack would be difficult.

Crack Growth Measurements

In transparent materials such as glass and many times in other ceramics, the position of the crack in a DT specimen can be determined optically through a traveling microscope or similar apparatus, similar to that carried out in DCB specimens. However, a significant advantage of the DT technique is its amenability to obtaining crack growth rates as a function of K_I without the need for the direct observation of the crack. Both a load relaxation and a constant displacement rate testing procedure (Evans 1972) have been shown to be effective in determining crack growth rates.

The displacement of the loading points, y, is directly related to the applied load, P, and the crack length, a, through

$$y = P(Ba + C) \tag{4.14}$$

where B and C are constants related to the elastic properties of the material, the dimensions of the specimen, and the test device. Taking the derivative of Eq. (4.15) with respect to time, t, leads to

$$\frac{dy}{dt} = (Ba + C)\frac{dP}{dt} + BPV \tag{4.15}$$

At constant displacement ($dy/dt = 0$)

$$V = -\frac{(Ba + C)}{BP}\frac{dP}{dt} \tag{4.16}$$

One can show also that

$$V = -\frac{P_i}{P^2}\left(a_i + \frac{C}{B}\right)\left(\frac{dP}{dt}\right) \tag{4.17}$$

where a_i is the initial crack length and P_i is the load at which the displacement of the crosshead of the testing machine is halted. For large a_i, Eq. (4.17) reduces to

$$V = -\frac{a_i P_i}{P^2}\left(\frac{dP}{dt}\right) \tag{4.18}$$

Crack growth data is collected by placing the precracked specimen in a machine capable of constant crosshead displacement. The load is raised to a predetermined value, P_i, at a relatively high rate of loading so that no crack growth occurs prior to reaching the fixed load. The crosshead is stopped, and the relaxation of the load is followed. Over 10 orders of magnitude in crack velocity have been reported using this procedure (Figure 4.8).

However, caution should be exercised in carrying out this technique to low crack velocities, i.e. below $10^{-7}\,\mathrm{m\,s^{-1}}$ because of possible interferences due to compliance variations in the test machine, temperature variations, or electrical perturbations.

A second method of acquiring crack growth data on the DT specimen is to load the specimen at a constant displacement rate (Evans 1972). Then from Eq. (4.15)

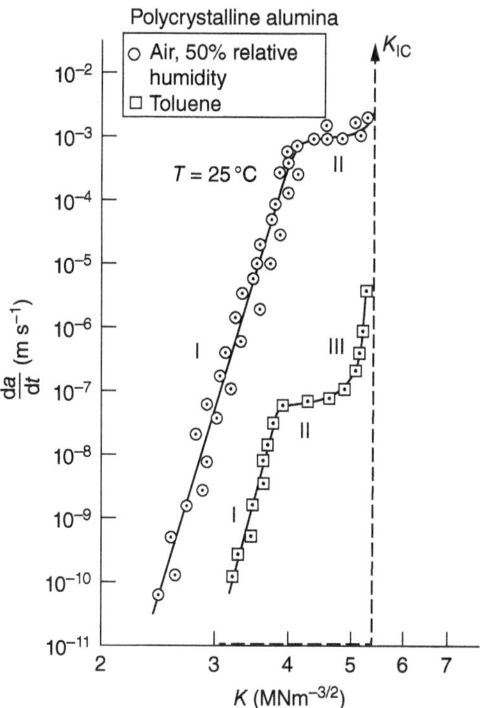

Figure 4.8 Crack growth data for a polycrystalline alumina using the load relaxation procedure. Source: After Evans (1972). Reproduced with permission of Springer Nature.

$$\frac{dy}{dt} = BPV = B'K_I V \tag{4.19}$$

A plateau in the load (Figure 4.9) is evidence that the crack is growing in a stable manner. The constant loading rate technique is particularly useful in obtaining data under severe environmental conditions in which either viewing the crack directly or depending on the stability of the system is questionable. One drawback to the use of the constant displacement rate method is that crack velocities that can be obtained are limited to $\gtrsim 10^{-5}\,\mathrm{m\,s^{-1}}$.

In summary, the DT technique is an excellent choice for obtaining crack growth data, particularly for materials where observing a crack optically is difficult, either because the material is opaque or because of the experimental conditions. It is less likely to be the choice for a critical fracture toughness determination for advanced ceramics primarily

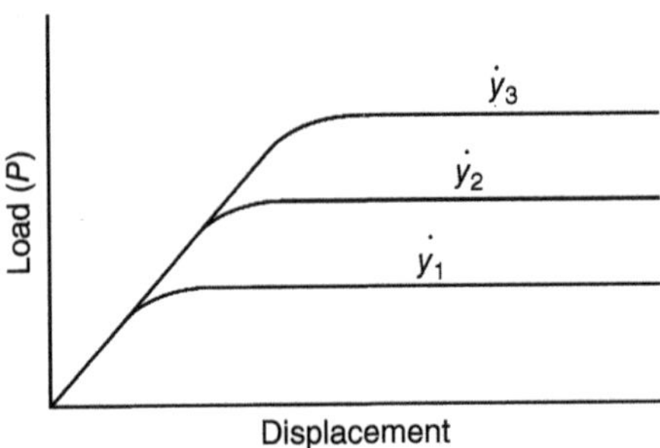

Figure 4.9 Plot of stress versus displacement rates for a double torsion specimen in which a crack is growing stably.

because of the amount of material required. However, the geologic and building material communities have made use of the ability to scale the specimen to large sizes to study rock and concrete fracture.

FLEXURAL TESTS

Background

A number of fracture mechanics test procedures are based on a rectangular bar (or beam) in which a notch or crack has been introduced and which is subsequently loaded to failure in flexure. The advantage of this test geometry is fairly obvious, simple geometry and loading. These tests include the single-edge-notch-bend (SENB) test, the chevron-notch-bend (CNB) test, the single-edge-precracked-beam (SEPB) method, the surface crack in flexure (SCF) test, and the work of fracture (WOF) test. Except for the WOF test, these procedures differ essentially only through the method by which a starting crack is introduced into the flexural bar.

Flexural Testing

All of the tests in this chapter require that a specimen be broken in flexure. Reproducible, accurate results of a flexural test depend on the specimen having the correct dimensions and machining and a correctly designed loading system. The particulars of how to conduct a flexural

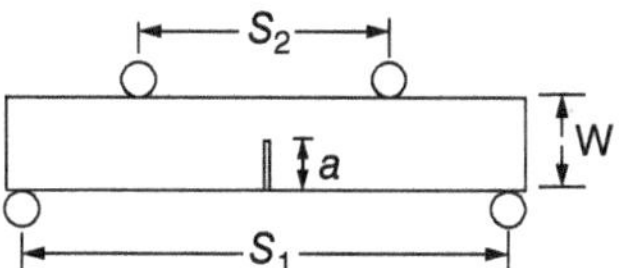

Figure 4.10 SENB specimen.

test for ceramics, including detailed descriptions of loading fixtures, are given in Chapter 5.

Single-edge-notch-bend Test

The first fracture mechanics test for brittle materials involving a flexural specimen was the SENB test (also called the notch-beam test) (Davidge and Tappin 1968). As its name suggests, it consists of a bar with a notch inserted in one of the narrow sides (Figure 4.10). The bar is loaded in either three- or four-point bending. It is typically assumed that in the case of brittle materials, a machining-induced crack exists at the root of the notch. This assumption is frequently incorrect and leads to overestimates of the fracture toughness. Because of serious questions regarding the nature of the initial crack, the SENB test remains dubious as a method that will yield accurate values of fracture toughness. Improvements in the test attained through better methods of introducing a true crack are discussed in the following sections.

Chevron-notch-bend Test

An improved version of the SENB test makes use of a flexural bar containing a symmetrical chevron notch (Figure 4.11). The specimen can be loaded in either three- or four-point bending. The presence of the chevron helps (but does not guarantee) the existence of an actual crack at the start of the test.

K_{IC} is calculated from the maximum load reached before failure using knowledge of the specimen compliance, thereby avoiding the need to determine an initial crack size. This specimen is one of three methods that have been adopted as an ASTM standard as designation C1421-01b, *Standard Test Methods for Determination of Fracture Toughness of Advanced Ceramics at Ambient Temperature*, as well as an ISO Standard 24370 (2005).

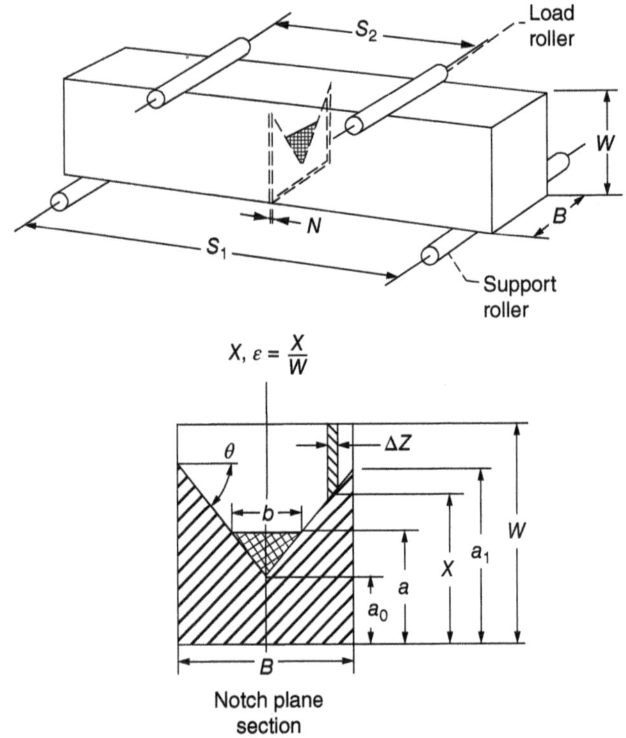

Figure 4.11 Schematic of CNBT specimen. Source: Munz et al. (1980). Reproduced with permission of Springer Nature.

Equation (4.20) is used to calculate fracture toughness, designated as K_{Ivb}. The subscript on K_{IC} in Eq. (4.20), as well as others noted in the ASTM standard, reflects the position that, because of factors such as a different response to environmentally enhanced crack growth, each of these tests may lead to a different value of fracture toughness:

$$K_{Ivb} = Y^*_{min} \left[\frac{P_{max} \left(S_o - S_i \right) 10^{-6}}{BW^{3/2}} \right] \tag{4.20}$$

In the ASTM standard, Y^*_{min} is defined as the minimum stress intensity factor coefficient and depends on the specific specimen geometry of the specimen. In order for a test to be valid, the force–displacement curve must exhibit a smooth maximum, as seen in Figure 4.12a. Curves

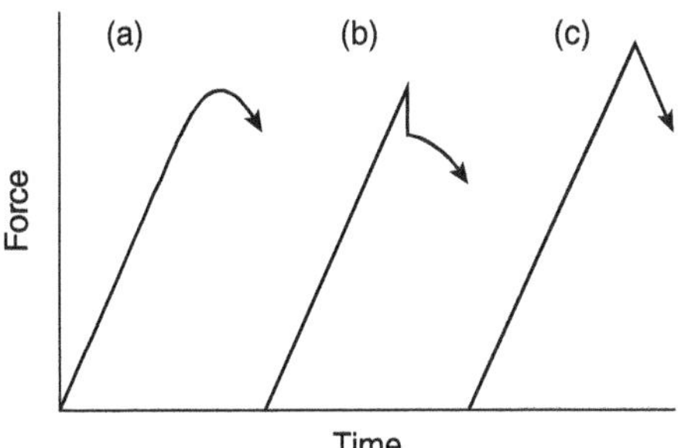

Figure 4.12 Acceptable (a and b) and (c) unacceptable force time traces.

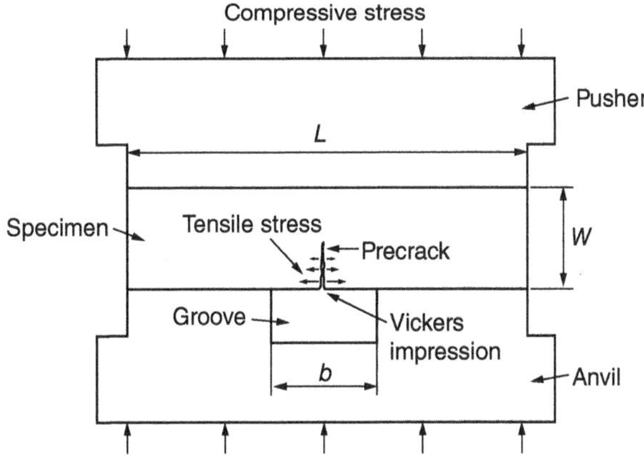

Figure 4.13 So-called "bridge indentation" device used to introduce a crack into an SEPB specimen. Source: After Nose and Fujii (1988). Reproduced with permission of John Wiley & Sons.

such as those shown in Figure 4.12b are indicative of a valid test as well, but care must be taken to use the P_{max} point on the diagram in the calculation.

Single-edge-precracked-beam

The SEPB test is one that involves the use of a specially designed system to introduce a controlled crack into the surface of a bend specimen (Figure 4.13).

This method now exists as both an ASTM and an ISO standard. The crack starter can be a machined notch, one or a series of Vickers indents or a Knoop indent with its long axis perpendicular to the longitudinal axis of the specimen. The precrack is then loaded in compression. A sound is created at pop-in, but it may be necessary to use a stethoscope to detect it. The use of an ultra-penetrating fluorescent dye penetrant helps to determine the final crack length before testing. As with other flexural tests, it is necessary to ensure that stable crack growth occurred, as evidenced from the shape of the force–displacement curve (Figure 4.12) in order to obtain an accurate value of K_{IC}. The particular expression used to calculate K_I depends on the width to depth ratio of the specimen and whether three- or four-point flexure is used (ISO Standard 15732:2003 2003; ASTM C1421-10 2010). This test is not used as often as others because of the complexity of the precracking procedure.

Surface Crack in Flexure

The SCF test involves placing a Knoop hardness indentation in the polished surface of a flexural bar, thereby forming a crack underneath the indentation site, then removing the residual stressed zone surrounding the indent, and fracturing the bar in four-point bending (Figure 4.14).

The SCF technique has become incorporated as an established procedure in the ASTM C1421-10 (2010) on fracture toughness as well as into an ISO Standard 18756:2003 (2003). In addition, a Standard Reference Material (2100) is available from NIST that can be used to calibrate the measurement.

The indentation force needed depends on the toughness of the material. Forces of 10–20 N are suitable for low toughness materials, e.g. glasses; 25–50 N for intermediate toughness materials, e.g. alumina; and 50–100 N for higher toughness materials, e.g. silicon nitride. In order to obtain an accurate value of fracture toughness by this technique, it

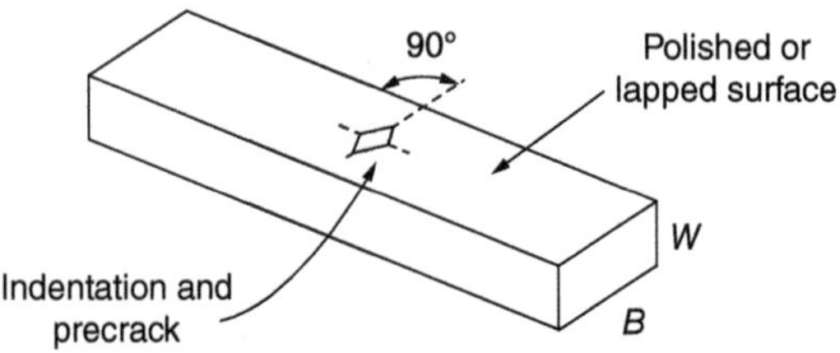

Figure 4.14 Schematic of SCF specimen.

is important that all of the stressed material is removed during polishing, e.g. to a depth of 4.5–5 times the depth of the indentation itself. Lateral cracks formed during the indentation process can also interfere with the attainment of an accurate value of K_{IC} (Quinn and Salem 2002). If the presence of lateral cracks is observed or suspected, 7–10 times the indentation depth should be removed.

After fracturing in four-point bending, the initial crack size is measured on the fracture surface. The ease and accuracy with which the crack can be measured is dependent on the microstructure of the material and to some degree the experience of the individual in fractographic analysis.

The fracture toughness is calculated from the following expression (ASTM C1421-10 2010):

$$K_{IC} = Y\left[\frac{3P_{max}(S_o - S_i)10^{-6}}{2BW^2}\right]\sqrt{a} \qquad (4.21)$$

where a is the crack depth, S_o and S_i are the outer and inner spans of the four-point flexure system, P_{max} is the maximum load recorded during the test, B and W are the width and depth of the specimen (see Figure 4.15) (the specimen can be tested in either orientation), and Y is the stress intensity coefficient that accounts for the flaw shape. The term in brackets is the flexural strength of the specimen.

As noted in Chapter 2, Y varies around the perimeter of the crack and is a complex function of the specimen dimensions. The largest value of Y should be used in the calculation.

The greater of Y_d and Y_s should be inserted into Eq. (4.20).

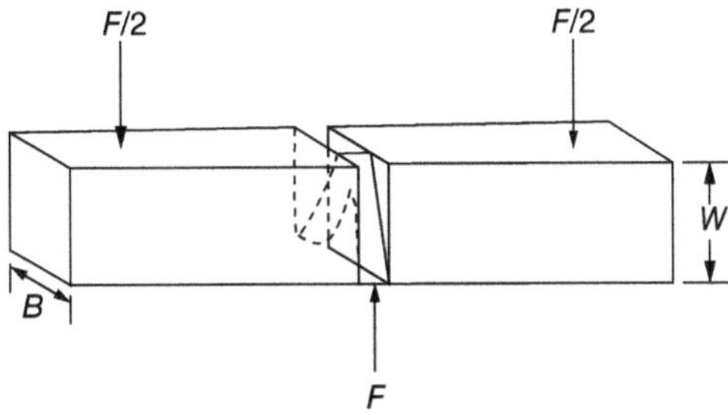

Figure 4.15 Typical work of fracture specimen. Source: After Nakayama (1965). Reproduced with permission of John Wiley & Sons.

Work of Fracture

The WOF test leads to a property quite different than that obtained from the previous test procedures in that cracks are growing stably rather than accelerating. Results from this test should not be compared directly with those obtained by other fracture mechanics tests.

The WOF test involves loading a bend bar containing a notch or a crack until the crack propagates through the entire specimen (Figure 4.16). The load trace is recorded, and the area under the loading curve, which is the energy absorbed in the fracture process, is measured. This absorbed energy per unit of surface formed during fracture is the fracture energy.

The WOF test measures not the critical fracture toughness but the average energy required to form the fracture surfaces as the crack extends through the depth of the specimen. Assuming a completely stiff loading system, the energy, U, stored in the specimen at the time a crack begins to propagate is given by

$$U = \frac{F^2 L^3}{96EI} \tag{4.22}$$

where F is the applied load at the point the crack starts to grow, E is Young's modulus, and I is the moment of inertia of the beam, $=bd^3/12$. If all of the applied energy is used exclusively to drive the crack, i.e. no kinetic energy is imparted to the halves of the bend bar, or other phenomena such as acoustic emission or heat generation occur, WOF can be calculated from the area under the load–displacement curve:

$$\gamma_{\mathrm{WOF}} = \frac{U}{2A_{\mathrm{p}}} \tag{4.23}$$

where A_{p}, the projected area of fracture, $=b(d-c)$, c being the initial crack or notch depth.

An essential assumption in this test is that the crack is growing stably and does not accelerate. That is, the energy needed to drive the crack should always be slightly less than that required for its growth. One must use a stiff test machine, i.e. one that stores a minimum of energy, and one must employ a specimen that requires a small load to initiate the crack so as not to store additional energy in the specimen that would be expended as heat, sound, or kinetic energy. This latter requirement suggests that starting cracks should be large with respect to the thickness of the specimen.

The shape of the load–displacement curve (Figure 4.12) is the primary indication of crack stability and therefore of the accuracy of the data. Recent evidence (Smith et al. 2009) suggests that the actual rather than the projected area should be used to calculate γ. This area can be measured using an atomic force microscope.

Although the values of γ reported for the WOF test are at times similar to those obtained by other procedures, this may be fortuitous. What is being measured is quite different.

DOUBLE CLEAVAGE DRILLED COMPRESSION

Background

The double cleavage drilled compression (DCDC) specimen is interesting because the crack grows smoothly through the center of the specimen without the need for guidance and remains stable because it extends into a decreasing K_I field. It has proved useful in studying crack growth at very low velocities and in the determination of adhesion between dissimilar materials.

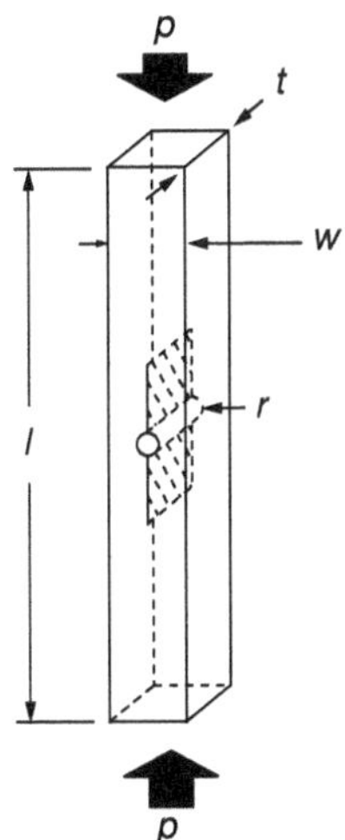

Figure 4.16 DCDC specimen.

Specimen

The specimen consists of a square rod, usually having approximate dimensions of 150×15×15 mm, with a circular, ~4 mm diameter, hole drilled through the center (as shown in Figure 4.16) (Janssen 1974). When loaded in compression, the presence of the hole modifies the applied compressive stress to create tensile stresses, leading to the formation of vertical cracks emanating from the hole.

The expression for G is given by

$$\frac{EG}{\pi\sigma^2 r\left(1-\upsilon^2\right)} = F\left(\frac{w}{r}\right)\left(\frac{a}{r}\right) \quad (4.24)$$

where E is Young's modulus, σ is the applied stress ($P/t\cdot w$), r is the hole radius, a is the crack length, and t and w are the specimen thickness and width, respectively, as shown in Figure 4.16.

The corresponding stress intensity factor is given by

$$\frac{\sigma\left(\pi r\right)^{1/2}}{K_I} = r + \left(\frac{0.235w}{r} - 0.259\right)\frac{a}{r} \quad (4.25)$$

Equation (4.25) is valid over the range $w/r \leq a/r \leq 15.2 \leq w/r \leq 4$. Because K_I is a decreasing function of crack length, what is measured as the crack propagates is the stress intensity for crack arrest rather than fracture toughness.

INDENTATION METHODS

Background

Indentation methods mean those procedures that involve using a hardness indentation coupled with the ensuing crack system as the primary source of the starting crack in a fracture mechanics test. There are two basic indentation test methods that have been proposed to determine K_{IC}: in the first, known as indentation fracture (IF), a Vickers indenter is used to create a hardness impression. The lengths of the cracks emanating from the corners of the indentation are measured, and this data is combined with knowledge of the elastic properties and hardness to calculate fracture toughness. In the second, called indentation strength (IS), the indentation is placed in a bar of the material, and the strength of the bar measured in flexure. This test differs from the SCF method discussed previously in this chapter in that the residual stress system associated with the indentation itself is not removed and is used as part of the driving force on the crack.

Indentation Fracture

In the IF technique, a Vickers hardness indentation is placed in the surface of the material producing cracks emanating from the corners of the hardness impression (Figure 4.17)

One of the serious complications and ultimate shortcomings of this test is the empirical nature of the expressions used to relate "K_{IC}" to the measurable parameters: load, hardness, elastic modulus, and crack length. Numerous expressions relating fracture toughness to indentation parameters have been suggested to calculate fracture toughness by this procedure (Ponton and Rawlings 1989a; Quinn and Bradt 2007). One of the more popular versions of such an expression is given in Eq. (4.26) (Anstis et al. 1981):

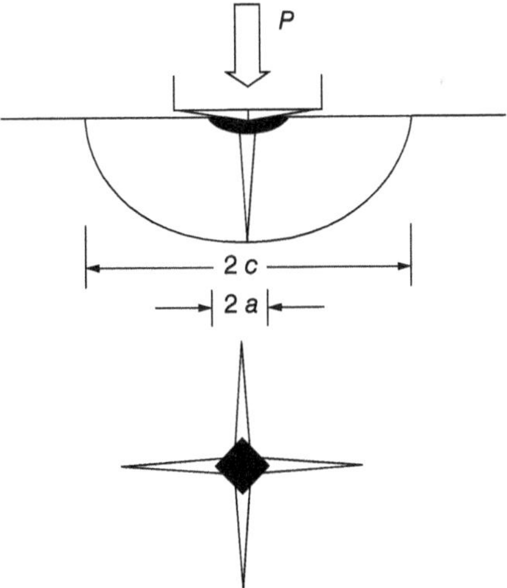

Figure 4.17 Crack pattern after indentation. Source: After Anstis et al. (1981). Reproduced with permission of John Wiley & Sons.

$$K_C = k\left(\frac{E}{H}\right)^{\frac{1}{2}} P c_0^{-\frac{3}{2}} \tag{4.26}$$

where E is Young's modulus, c_0 is the measured crack length, P is the indentation load, k is claimed to be a material-independent constant, and H is the measured hardness determined from the expression

$$H = \frac{P}{\alpha_0 a^2} \tag{4.27}$$

In this expression, a is the size of the indentation impression, and α_0 is a numerical constant. The term K_C rather than K_{IC} is used to emphasize that the procedure should not be considered to lead to a "standard" fracture toughness. Several significant concerns have been raised regarding this test procedure (Quinn and Bradt 2007).

- This method cannot yield a value of fracture toughness that is consistent with the definition of K_{IC} given in Chapter 1, namely, the stress intensity at which unstable crack growth occurs. Rather than accelerating, the crack decelerates and stops. This technique would actually yield a value of K_I at which crack arrest takes place. There is no basis for assuming that this crack arrest K_I corresponds to a fracture toughness value obtained by accepted methods in which a single crack is grown to failure.
- The crack system itself can be extremely complex depending on the microstructure of the material. More than one crack can be present, and the presence of lateral cracks that extend more or less parallel to the surface below the indentation can further interfere with the process.
- None of the expressions used to calculate K_C are fundamental in nature. There are adjustable constants that are not material independent.

Could this test be used to rank materials without depending on the numerical accuracy of K_C? This too is questionable. Ponton and Rawlings (1989b) compared materials using a number of the expressions proposed to calculate fracture toughness and concluded that an accurate ranking of materials with respect to fracture toughness depended on the set of materials chosen and the expression used to calculate K_C.

Nonetheless, there are situations in which there are no other techniques available to determine fracture toughness. A particular example is the determination of the fracture resistance of small single crystals, where the size of the available crystals (of the order of a few millimeters or less) precludes the use of standard fracture mechanics tests. Determination of the fracture toughness of single crystals poses an additional problem, however, due to the anisotropic nature of the elastic properties. It is surprising therefore to find that in many cases values of K_{IC} for single crystals determined by the IF method compare reasonably well with those measured by other techniques. As an example, the IF method has been used to determine the fracture toughness of high T_C superconducting crystals, which were $<1\ mm^2$ in area (Raynes et al. 1991).

Another use for the IF test is in the investigation of the possible role of environment, e.g. water, on crack extension. It is well known that if environmentally enhanced crack growth takes place in a particular material, then the cracks will extend to a greater distance than C_0, the crack length measured in an inert environment. This factor must be taken into account in any calculation of fracture toughness, but it has been shown that it also possible to use the ratio of crack lengths in various environments to obtain at least a semiquantitative measure of the material's susceptibility to subcritical crack growth (Gupta and Jubb 1981; White et al. 1991).

Using the expression

$$V = V_0\left(\frac{K_I}{K_{IC}}\right)^N \tag{4.28}$$

where V_0 and N are empirical material parameters, leads to the following expression for the ratio of crack lengths in two different environments:

$$\frac{c}{c_0} = \left[(3N+2)\frac{V_0 t}{2c_0}\right]^{2/(3N+2)} \tag{4.29}$$

where c_0 is the initial crack length after indentation and c is the crack length measured at time, t. This expression is valid when

$$(3N+2)\frac{V_0 t}{2c_0} \gg 1$$

Despite the fact that some data is lost because of the difficulty in obtaining measurements at very short times, this procedure can be used to determine values of N that agree reasonably well with those obtained from classical fracture mechanics specimens.

This indentation procedure has also been employed as a way of determining effects of residual stresses and applied electric fields on crack growth. In fact, it has become a primary technique for studying fracture in piezoelectric materials.

Indentation Strength

The IS technique to determine fracture toughness was suggested by Chantikul et al. (1981) and is based on the use of an indentation-induced surface crack as the starting point for a test. Instead of removing the stress field due to the indentation, it is combined with the far field stress as part of the driving force on the crack. Fracture strength can be measured in either uniaxial or biaxial flexure.

Chantikul et al. (1981) derived the following expression for K_c:

$$K_c = \eta_v^R \left(\frac{E}{H}\right)^{1/8} \left(\sigma P^{1/3}\right)^{3/4} \tag{4.30}$$

where E is Young's modulus, H is the hardness, σ is the fracture strength, P is the indentation load, and η_v^R is a geometrical constant given by

$$\eta_v^R = \left[\left(\frac{256}{27}\right)(\pi\Omega)^{3/2} k\right] \tag{4.31}$$

where Ω is a crack geometry factor and k is the same constant for Vickers-produced radial cracks as given in Eq. (4.26).

An average value for $\eta_v^R = 0.59$ was calculated by using K_C values obtained from "reference" ceramics by conventional fracture mechanics tests (Chantikul et al. 1981) (Figure 4.18).

Chantikul et al. claim an accuracy of 30–40% for K_C, but one should note that some materials, glasses in particular, deviate from the trend

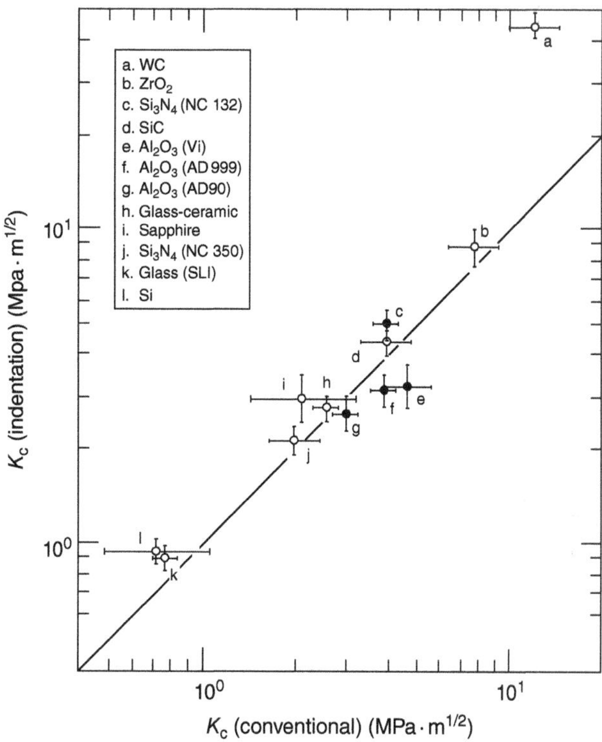

Figure 4.18 Fracture toughness calculated using the indentation-strength method compared with that determined from more conventional fracture mechanics methods (Chantikul et al. 1981).

line, and caution must be exercised in assuming that fracture toughness values for glasses are within this range. There are other factors that can lead to inaccuracy in the determination of K_C by this method. These include interactions between the radial and lateral cracks and microstructural effects, including anisotropy. At small indentation loads, one must make sure that failure began from the indentation site and not from a preexisting flaw in the surface or edge of the specimen.

One way to determine whether this method is applicable to a particular material is to logarithmically plot fracture strength as a function of indentation load. The slope of the line should be 1/3. Any deviation from this slope indicates the presence of residual stresses in the material that can interfere with the measurement. In addition, as

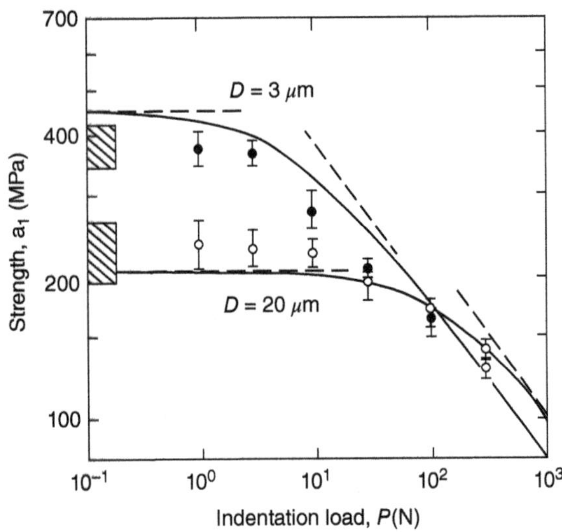

Figure 4.19 Strength-indentation load plot for a glass ceramic showing a deviation from the $P^{⅓}$ slope expected from Eq. (4.30).

seen in Figure 4.19, the data can exhibit plateaus in strength. One must take care to ensure that failure began from the indentation site and not from "natural flaws" in the material. Under no circumstances should a value of K_C be taken from this regime. On the other hand, such deviations can be valuable in obtaining information regarding the influence of microstructural-level stresses on fracture.

Toughness Evaluation from Edge Chipping

In many cases, loading near free surfaces, i.e. edges, will result in crack formation known as chipping. This chipping phenomenon has been known for centuries. Prehistoric humans took advantage of this phenomenon to make tools and tips of weapons using flint knapping (Quinn et al. 2000). Quinn et al. (2010) found that edge chipping can be used to determine the resistance to chipping in porcelain fused to metal (PFM) and veneered zirconia specimens, both materials used for dental crowns. They found that the two materials behaved similarly in the resistance to edge chipping even though the hardness of the substrates was quite different. Morrell and Gant (2001) demonstrated the use of edge chipping in hard materials and tool steels. They found a relationship between "edge toughness" and the distance from the indent and the edge of the specimen. Quinn et al. (2000) found a correlation between a term they called "edge toughness" and the fracture toughness for a number of materials. Their work agrees in principle with the work of Chai and Lawn (2007) that will be discussed below. Gogotsi and Mudrik (2010) and Gogotsi (2013) found a correlation between the fracture toughness and the critical load to fracture using a conical indenter to obtain edge chipping in a number of materials including glass and Y-TZP. He concluded that more work needs to be done to get a theoretical basis for the correlation.

The mechanics of the formation is extremely difficult to model. However, an observation by Chai and Lawn (2007) and Chai et al. (2011) noted that there is a similarity in the chip formation that allows a fracture mechanics relationship to be used relating the maximum load, P_F, during chip formation to the size of the chip, location from the free surface, h, and the toughness, K_C, of the material (Figure 4.20):

$$K_C = \frac{P_F}{\beta} h^{3/2} \tag{4.32}$$

where β is a geometric, material-independent constant ($\beta = 9.3 \pm 1.3$).

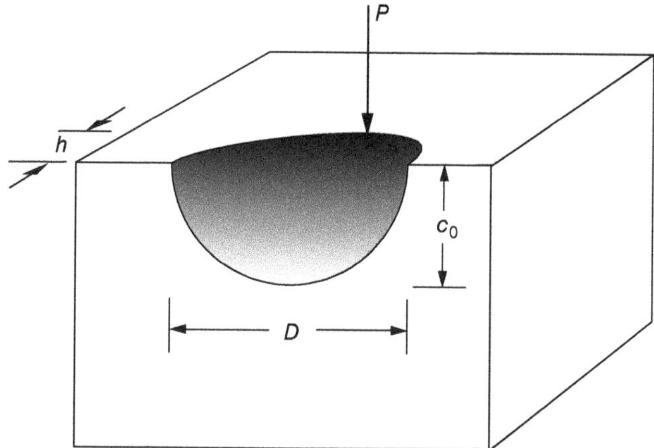

Figure 4.20 Schematic of edge chip with dimensions. P is the indent load, h is the distance from the edge, c_0 is the crack depth, and D is the crack width. Source: After Chai (2014). Reproduced with permission of Elsevier.

There are several assumptions and conditions in the analysis that must be recognized when using this approach to toughness due to edge chipping. The theory is based on the principle of geometric similarity and the existence of a Boussinesq stress field associated with indentation. The theory assumes no slow crack growth, no significant R-curve behavior, and monotonic loading that is normal to the surface and that two conditions are met, i.e. that median cracks initiate at low loads and that the cracks extend beyond the plastic contact zone. These conditions are observed for many applications. Thus, this may be a very useful technique for applications in dentistry, for example. Chai et al. (2011) demonstrated how this theory may be applied to the analysis for edge fracture in teeth. They were able to use Eq. (4.32) to estimate chipping loads and fracture toughness in teeth, even with the complex geometry of the cusp and complicated microstructure of enamel.

Toughness Evaluation from Nanoindentation

Although the nanoindenter has mostly been used for determination of nano-hardness and an evaluation of the local elastic modulus, it can also be used for fracture resistance determination. Oyen and Cook (2009) present a practical guide for nanoindentation, especially for biomaterials and biological materials. Various options for test methods are presented in the paper. They show that the toughness of tooth enamel as well as other materials can be measured using nanoindentation. They base their work on the indentation work of Anstis et al. (1981) discussed above with the addition of a calibration constant for nanoindentation, i.e. using Eq. (4.26) with $k = 62.5$.

MIXED-MODE TESTS

There are several techniques that have been used to assess the effect of mixed-mode loading on cracks. The most popular because of the relative ease of introduction of the initial crack is the flexure specimen (cf. Figure 2.5). Indentation cracks have been used to introduce controlled cracks using either the Knoop or Vickers diamond indenters. The process of indentation is relatively straightforward. However, care must be taken in aligning the cracks at specific angles to the applied stress. Petrovic and Mendiratta (1976, 1977) and Petrovic (1985) have examined mixed-mode fracture toughness in hot pressed silicon nitride using surface flaws produced by Knoop indentation in a four-point bending test. Khandelwal et al. (1995) have found the high temperature mixed-mode fracture toughness of hot isostically pressed PY6 silicon nitride using bend bars that contained crack indentations produced using the Vickers hardness indenter and tested using the four-point bend test. Straight-through cracks have also been used in flexure specimens (Suresh et al. 1990). It is best to perform all these tests in an inert environment, if possible. The effects of slow crack growth for mixed-mode loading have been discussed in Chapter 2. Numerous investigators have used the flexure specimen in assessing the effects of mixed-mode loading on crack propagation: Marshall (1984), Glaesemann et al. (1987), and Petrovic and Mendiratta (1976, 1977). There also has been limited mixed-mode testing using tensile and torsion rods (Suresh and Tschegg 1987).

The Brazilian disk (BD) test is commonly used to measure the tensile strength and fracture toughness of brittle materials like rocks, concrete, and ceramics. The major advantage of this test is that the test can be done for a range of mode mixities, in principle, ranging from pure mode I to pure mode II. A centrally cracked BD specimen is a very commonly used geometry for mixed-mode fracture studies on ceramics. The BD test was first proposed by Carniero and Barcellos (1953). This test is commonly used to measure the tensile strength and fracture toughness of brittle materials like rocks, concrete, and ceramics. Zhou et al. (2006) have even used a BD specimen of polymethylmethacrylate (PMMA) with a chevron notch to determine its mode I fracture toughness. The test basically has a circular disk loaded in compression, and tensile stress acting at the center (perpendicular to the loading direction) splits the disk along the loaded diameter. The advantage of this test is that the test can be performed under a range of mode mixities, in principle, ranging from pure mode I to pure mode II and, thus, the fracture toughness in tension, shear, and in any combination of tension and shear can be evaluated. However, pure mode II is difficult to achieve because of crack closure due to compressive loading. Simple specimen geometry, minimal fixture requirements, and its potential for use at higher temperatures are other major merits of this test (Shetty et al. 1985, 1986, 1987). BD specimens can also be used to measure interfacial fracture toughness in biomaterials. Tong et al. (2007) used the BD test to determine the interfacial fracture toughness in bone–cement interfaces. Most of the mixed-mode fracture studies are done by producing precracks in the samples. There are two approaches for generating precracks in a specimen to determine its fracture toughness. Precracks can be formed from surface flaws produced by indentations. Larger precracks can be introduced in the ceramic specimens using the chevron notch or a sharp-edged notch for which there is stable crack growth during the initial precracking. There have been quite a number of studies conducted on mixed-mode fracture of ceramics by using either of these approaches. Awaji and Sato (1978) employed this test for examining the fracture toughness of graphite, plaster, and marble in mixed-mode loading using machined central cracks.

There are many theories that have been developed to describe the fracture criteria in mixed-mode loading. Many of these theories were discussed in Chapter 2. A biaxial stress state is generated in the specimen under diametral compression, as shown in Figure 4.21. The solutions for the full biaxial stress field generated in a thin isotropic disk are given by Mitchell (1961):

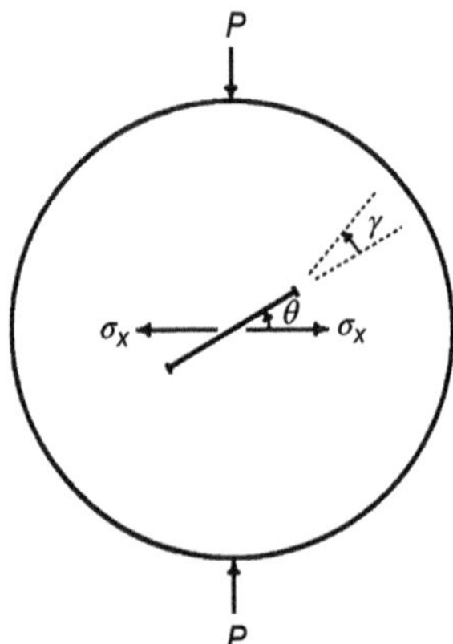

Figure 4.21 Schematic of the diametral compression test. The maximum tensile stress is at the center of the specimen. This specimen can be used to determine the effect of mixed-mode loading. When the crack is perpendicular to the loading is pure mode I loading. This specimen is also called the Brazilian disk specimen.

$$\sigma_x = -\frac{2P}{\pi B}\left[\frac{(R-y)x^2}{r_1^4} + \frac{(R+y)x^2}{r_2^4} - \frac{1}{2R}\right] \tag{4.33}$$

$$\sigma_y = -\frac{2P}{\pi B}\left[\frac{(R-y)^3}{r_1^4} + \frac{(R+y)^3}{r_2^4} - \frac{1}{2R}\right] \tag{4.34}$$

$$\tau_{xy} = \frac{2P}{\pi B}\left[\frac{(R-y)^2 x}{r_1^4} - \frac{(R+y)^2 x}{r_2^4}\right] \tag{4.35}$$

$$r_1^2 = (R-y)^2 + x^2 \tag{4.36}$$

$$r_2^2 = (R+y)^2 + x^2 \tag{4.37}$$

where P is the load applied, D is the diameter or the disk specimen, and B is the thickness of the disk specimen. At the center of the disk,

$$\sigma_x(0,0) = \frac{2P}{\pi DB} \tag{4.38}$$

$$\sigma_y(0,0) = -\frac{6P}{\pi DB} \tag{4.39}$$

$$\tau_{xy}(0,0) = 0 \tag{4.40}$$

It can be seen from Eq. (4.38) that $\sigma x(0,0)$ is the maximum tensile stress and this value of the stress remains approximately the same for a significant region near to the center of the disk. The indented flaws can be approximated as semicircular flaws, and the stress intensity factors for such flaws, loaded under mixed-mode conditions, are given by

$$K_I = Y_1 \sigma \sqrt{c} \sin^2 \theta \tag{4.41}$$

$$K_{II} = Y_2 \sigma \sqrt{c} \sin\theta \cos\theta \tag{4.42}$$

where σ is the applied stress, c is the equivalent semicircular crack length, and θ is the angle between the initial crack plane and the plane in which the crack propagates. c is given by $\sqrt{ab}$ where a is the length of the semiminor axis of the critical flaw and $2b$ is the length of the semimajor axis of the elliptical crack. Y_1 is a geometric constant equal to 1.68 for indented specimens (Marshall et al. 1980). Kassir and Sih (1966) developed the expression for K_{II} for penny-shaped cracks with the geometric constant Y_2 given by

$$Y_2 = \frac{4}{\sqrt{\pi}\left(2-\nu\right)} \tag{4.43}$$

The value of Y_2 is 1.29 for a Poisson ratio of $\nu = 0.25$. Thus, the critical stress intensity factors can be calculated from the measurement of the flaw dimensions.

The combination of K_I and K_{II} should be performed based on the assumptions of the appropriate theory of mixed-mode loading. For example, the coplanar, maximum strain energy release rate criterion results in

$$K_c = \left(K_I^2 + K_{II}^2\right)^{1/2} \tag{4.44}$$

where K_c is the critical stress intensity factor at crack propagation for the mixed-mode condition. The appropriateness of using other criteria is discussed in Chapter 2.

SUMMARY

The salient points with respect to each of the fracture mechanics techniques discussed in this chapter are summarized below.

Double Cantilever Beam

Fracture toughness can be determined using many of the DCB modifications, but no standard exists. K_{IC} values measured on DCB specimens are comparable with those obtained by other methods. Because of the crack-length independence of K_I, the AMDCB variation is especially recommended for obtaining crack propagation data. The DCB configuration is particularly useful for obtaining crack growth data, but generally the crack length must be measured optically. For translucent or opaque specimens, proper lighting is a key to observing crack motion. In polycrystalline materials the crack appears to grow sporadically, and there can be significant scatter to the data. The specimen must contain a groove approximately one-half of the specimen thickness in order to guide the crack. This groove can cause complications due to machining damage. Attaching a loading system to a specimen can be difficult especially for small specimens.

Double Torsion

K_I is crack-length independent. All loading is in compression. Crack growth data can be obtained without the need for direct observation of the crack, making this configuration useful for testing at elevated temperatures and in harsh environments. These methods include:

- Load relaxation – Care must be taken to avoid drift in the loading system or data recorder due to electronic or temperature variations.
- Load plateau – Limited to higher crack velocities.

This can be a very useful test for materials such as concrete where large-size specimens can be prepared, but for advanced or experimental materials, size and material use requirements are a significant drawback.

Bend Tests

Bend tests involve a simple specimen and loading geometry, are relatively inexpensive, but cannot be used for monitoring crack growth. Variations include:

- *SENB* – The data is highly questionable unless one can guarantee the presence of an actual crack at the root of the notch.
- *CNB* – ASTM and ISO standards exist.
- *SEPB* – ASTM and ISO standards exist. Precracking can be a tedious operation.

- *SCF* – ASTM and ISO standards exist. Crack size can be adjusted to grain size to resemble actual surface flaws. Some skill required to measure the initial crack size on the fracture surface.
- *WOF* – This test yields an *average* fracture energy rather than a *critical* fracture toughness.

Double Cleavage Drilled Compression Test

This test does not yield true fracture toughness, but the K_I at which crack arrest takes place. This test configuration lends itself to measurement of crack growth resistance along an interface. For most materials having reasonably large fracture toughness, the loads required to propagate a crack can be large, and buckling of the column can become a problem.

Indentation Tests

Indentation tests whether they are IF, IS, or edge chipping should not be assumed to yield an accurate value of K_{IC}. These tests can, however, be used with caution to ascertain a relative resistance to crack growth. Indentation tests can also yield a measure of the susceptibility of a material to environmentally enhanced crack growth.

Mixed-mode Tests

The most important aspects for mixed-mode loading are the introduction of the starter crack and eliminating any effects of slow crack growth before or during testing. If the starter crack is introduced by indentation then any residual stress introduced by the indentation process should be either eliminated or accounted for. The other important aspect is the selection of the method of combining K_I and K_{II}. It is not clear which is the best method. However, recent results suggest that the modified strain energy density method is optimum for many materials.

QUESTIONS

1. What is (are) the advantage(s) of selecting the constant moment double cantilever beam (CMDCB) specimen over the double cantilever beam (DCB) specimen?

2. If you need to test a polycrystalline ceramic with a fine grain structure to determine the toughness at elevated temperature, which specimen configuration would you select? Why?
3. If you need to test a polycrystalline ceramic with a fine grain structure to determine the stress corrosion susceptibility at elevated temperature, which specimen configuration would you select? Why?
4. When would you select a flexure beam configuration for determining fracture toughness as opposed to selecting double cantilever beam or double torsion specimens?
5. What are the advantages and disadvantages of the different methods of introducing a controlled crack into a flexure specimen for determining the fracture toughness?
6. Discuss the advantages and disadvantages of determining the fracture toughness using indentation techniques.

REFERENCES

Anstis, G.R., Chantikul, P., Lawn, B.R., and Marshall, D.B. (1981). A critical evaluation of indentation techniques for measuring fracture toughness: I, direct crack measurements. *J. Am. Ceram. Soc.* 64: 533–538.

ASTM C1421-10 (2010). *Standard Test Methods for Determination of Fracture Toughness of Advanced Ceramics at Ambient Temperature*. West Conshohocken, PA: ASTM International.

Awaji, H. and Sato, S. (1978). Combined mode fracture toughness measurement by the disk test. *J. Eng. Mater. Technol.* 100: 175–182.

Carniero, F.L.L.B. and Barcellos, A. (1953). Résistance à la traction des bétons. *Bull. RILEM (I)* 13: 97–108. (re-edited in 1997 by RILEM in RILEM fifty years of evolution of science and technology of building materials and structures, ed. F. Wittmann, Aedificatio, 91–120).

Chai, H. (2014). On the mechanical properties of tooth enamel under spherical indentation. *Acta Biomater.* 10: 4852–4860.

Chai, H. and Lawn, B.R. (2007). A universal relation for edge chipping from sharp contacts in brittle materials: a simple means of toughness evaluation. *Acta Mater.* 55: 2555–2561.

Chai, H., Lee, J.J.W., and Lawn, B.R. (2011). On the chipping and splitting of teeth. *J. Mech. Behav. Biomed. Mater.* 4: 315–321.

Chantikul, P., Anstis, G.R., Lawn, B.R., and Marshall, D.B. (1981). A critical evaluation of indentation techniques for measuring fracture toughness: II, strength method. *J. Am. Ceram. Soc.* 64: 539–543.

Davidge, R.W. and Tappin, G. (1968). The effective surface energy of brittle materials. *J. Mater. Sci.* 3: 165–173.

Evans, A.G. (1972). Method for evaluating the time-dependent failure characteristics of brittle materials and its application to polycrystalline alumina. *J. Mater. Sci.* 7: 1137–1146.

Evans, A.G. (1974). Fracture mechanics determinations. In: *Fracture Mechanics of Ceramics*, vol. 1 (ed. R.C. Bradt, D.P.H. Hasselman and F.F. Lange), 17–24. New York: Plenum Publishing Co.

Freiman, S.W., Mulville, D.R., and Mast, P.W. (1973). Crack propagation studies in brittle materials. *J. Mater. Sci.* 8: 1527–1533.

Freiman, S.W., McKinney, K.R., and Smith, H.L. (1974). Slow crack growth in polycrystalline ceramics. In: *Fracture Mechanics of Ceramics*, vol. 3 (ed. R.C. Bradt, D.P.H. Hasselman and F.F. Lange), 659. Plenum Publishing Co.: New York.

Fuller, E.R. Jr. (1979). *Fracture Mechanics Applied to Brittle Materials*, American Society for Testing and Materials, STP 678 (ed. S.W. Freiman), 3–18. Philadelphia, PA: ASTM.

Glaesemann, S.G., Ritter, J.E. Jr., and Jakus, K. (1987). Mixed mode fracture in soda lime glass using indentation flaws. *J. Am. Ceram. Soc.* 70 (9): 630–636.

Gogotsi, G.A. (2013). Edge chipping resistance of ceramics: problems of test method. *J. Adv. Ceram.* 2: 370.

Gogotsi, G.A. and Mudrik, S.P. (2010). Fracture resistance of technical and optical glasses: edge flaking of specimens. *Strength Mater.* 42, No. 3: 280–286.

Gupta, P.K. and Jubb, N.J. (1981). Post-indentation slow growth of radial cracks in glasses. *J. Am. Ceram. Soc.* 64: C-112–C-114.

ISO Standard 15732:2003 (2003). *Fine Ceramics (Advanced Ceramics, Advanced Technical Ceramics) – Test Method for Fracture Toughness of Monolithic Ceramics at Room Temperature by Single Edge Precracked Beam (SEPB) Method.* Geneva, Switzerland: ISO.

ISO Standard 18756:2003 (2003). *Fine Ceramics (Advanced Ceramics, Advanced Technical Ceramics) – Determination of Fracture Toughness of Monolithic Ceramics at Room Temperature by the Surface Crack in Flexure (SCF) Method.* Geneva, Switzerland: ISO.

ISO Standard 24370 (2005). *Fine Ceramics (Advanced Ceramics, Advanced Technical Ceramics) – Test Method for Fracture Toughness of Monolithic Ceramics at Room Temperature by Chevron-notched Beam (CNB) Method.* Geneva, Switzerland: ISO.

Janssen, C. (1974). Specimen for fracture mechanics studies on glass. In: *Proceedings of 10th International Congress on Glass*, 10.23–10.30. Tokyo: Ceramic Society of Japan.

Kassir, M.K. and Sih, G.C. (1966). Three-dimensional stress distribution around an elliptical crack under arbitrary loadings. *J. Appl. Mech. Trans. ASME Ser. E* 33 (3): 601–611.

Khandelwal, P., Majumdar, B.S., and Rosenfield, A.R. (1995). Mixed mode high temperature toughness of silicon nitride. *J. Mater. Sci.* 30: 395–398.

Marshall, D.B. (1984). Mechanisms of failure from surface flaws in mixed-mode loading. *J. Am. Ceram. Soc.* 67: 110–116.

Marshall, D.B., Lawn, B.R., and Mecholsky, J.J. (1980). Effect of residual contact stresses on mirror/flaw size relation. *J. Am. Ceram. Soc.* 63 (5–6): 358–360.

Mitchell, N.B. (1961). The indirect tension test for concrete. *Mater. Res. Stand.* 1: 780–788.

Morrell, R. and Gant, A.J. (2001). Edge chipping of hard materials. *Int. J. Refract. Met. Hard Mater.* 19: 293–301.

Mostovoy, S., Crosley, P.B., and Ripling, E.J. (1967). Use of crack-line-loaded specimens for measuring plane-strain fracture toughness. *J. Mater.* 2: 661–681.

Munz, D., Bubsey, R.T., and Srawley, J.E. (1980). Compliance and stress intensity coefficients for short bar specimens with chevron notches. *Int. J. Fract.* 16: 359–374.

Nakayama, J. (1965). Direct measurement of fracture energies of brittle heterogeneous materials. *J. Am. Ceram. Soc.* 48: 583–587.

Nose, T. and Fujii, T. (1988). Evaluation of fracture toughness for ceramic materials by a single-edge-precracked-beam method. *J. Am. Ceram. Soc.* 71: 328–333.

Outwater, J.O., Murphy, M.C., Kumble, R.G., and Berry, J.T. (1974). *Fracture Toughness and Slow Stable Cracking*, American Society for Testing and Materials, STP 559, 127–138. Philadelphia, PA: ASTM.

Oyen, M.L. and Cook, R.F. (2009). A practical guide for analysis of nanoindentation data. *J. Mech. Behav. Biomed. Mater.* 2: 396–407.

Petrovic, J.J. (1985). Mixed-mode fracture of hot-pressed Si_3N_4. *J. Am. Ceram. Soc.* 68: 348–355.

Petrovic, J.J. and Mendiratta, M.G. (1976). Mixed mode fracture from controlled surface flaws in hot pressed silicon nitride. *J. Am. Ceram. Soc.* 59 (3–4): 163–167.

Petrovic, J.J. and Mendiratta, M.G. (1977). Correction of 'mixed-mode fracture from controlled surface flaws in hot-pressed Si_3N_4. *J. Am. Ceram. Soc.* 60 (9–10): 463.

Ponton, C.B. and Rawlings, R.D. (1989a). Vickers indentation fracture toughness test, part 1, review of literature and formulation of standardised indentation toughness equations. *Mat. Sci. Technol.* 5: 865–872.

Ponton, C.B. and Rawlings, R.D. (1989b). Vickers indentation fracture toughness test, part 2, application and critical evaluation of standardized indentation toughness equations. *Mat. Sci. Technol.* 5: 961–976.

Quinn, G.D. and Bradt, R.C. (2007). On the Vickers indentation fracture toughness test. *J. Am. Ceram. Soc.* 90: 673–680.

Quinn, G.D. and Salem, J.A. (2002). Effect of lateral cracks on fracture toughness determined by the surface-crack-in-flexure method. *J. Am. Ceram. Soc.* 85: 873–880.

Quinn, J., Su, L., Flanders, L., and Lloyd, I. (2000). Edge toughness and material properties related to the machining of dental ceramics. *Machining Sci. Technol.* 4 (2): 291–304.

Quinn, J.B., Sundar, V., Parry, E.E., and Quinn, G.D. (2010). Comparison of edge chipping resistance of PFM and veneered zirconia specimens. *Dent. Mater.* 26: 13–20.

Raynes, A.S., Freiman, S.W., Gayle, F.W., and Kaiser, D.L. (1991). Fracture toughness of $YBa_2Cu_3O_{6+}\delta$ single crystals: anisotropy and twinning effects. *J. Appl. Phys.* 70: 10–14.

Shetty, D.K., Rosenfield, A.R., and Duckworth, W.H. (1985). Fracture toughness of ceramics measured by a chevron notch diametral compression test. *J. Am. Ceram. Soc.* 68 (12): C325–C327.

Shetty, D.K., Rosenfield, A.R., and Duckworth, W.H. (1986). Mixed mode fracture of ceramics in diametral compression. *J. Am. Ceram. Soc.* 69 (6): 437–443.

Shetty, D.K., Rosenfield, A.R., and Duckworth, W.H. (1987). Mixed mode fracture in biaxial stress state: application of the Brazilian disk test. *Eng. Fract. Mech.* 26 (6): 1825–1840.

Shyam, A. and Lara-Curzio, E. (2006). The double-torsion testing technique for determination of fracture toughness and slow crack growth behavior of materials: a review. *J. Mater. Sci.* 41: 4093–4104.

Smith, R.L. III, Mecholsky, J.J. Jr., and Freiman, S.W. (2009). Estimation of fracture energy from the work of fracture and fracture surface area: I. Stable crack growth. *Int. J. Fract.* 156: 97–102.

Suresh, S. and Tschegg, E.K. (1987). Combined mode I-mode 111 fracture of fatigue-precracked alumina. *J. Am. Ceram. Soc.* 70 (10): 726–733.

Suresh, S., Shih, C.F., Morrone, A., and O'Dowd, N.P. (1990). Mixed-mode fracture toughness of ceramic materials. *J. Am. Ceram. Soc.* 73 (5): 1257–1267.

Tong, J., Wong, K.Y., and Lupton, C. (2007). Determination of interfacial fracture toughness of bone-cement interface using sandwich Brazilian disks. *Eng. Fract. Mech.* 74 (12): 1904–1916.

White, G.S., and Wilson, A.M. (1991). Indentation determination of crack growth parameters in gallium arsenide. *J. Am. Ceram. Soc.* 74: 419–421.

Wiederhorn, S.M., Shorb, A.M., and Moses, R.L. (1968). Critical analysis of the theory of the double cantilever method of measuring fracture-surface energies. *J. Appl. Phys.* 39: 1569–1572.

Wu, C.C., McKinney, K.R., and Lewis, D. (1984). Grooving and off-center crack effects on applied-monent double-cantilever-beam tests. *J. Am. Ceram. Soc.* 67: C-166–C-168.

Zhou, J., Wang, Y., and Xia, Y. (2006). Mode-I fracture toughness measurement of PMMA with the Brazilian disk test. *J. Mater. Sci.* 41 (17): 5778–5781.

CHAPTER 5

Strength Testing

INTRODUCTION

The strength of a material tells us what we need to know about the loads that a structure or component made from a brittle material will sustain. The main consideration is that their strength depends on the flaw severity and stress state at the origin of failure and thus is statistical in nature. Stresses can arise, not just to externally applied mechanical loads but also due to processing-induced stresses, thermal gradients, phase transformations, or the presence of electric fields (Pohanka et al. 1978). The flaw severity and location, coupled with the material's resistance to crack extension, i.e. its fracture toughness, are the key parameters that control failure (Wachtman 1996). Because there are no effective nondestructive techniques for brittle materials to identify flaw sizes prior to fracture, strength is the only measure of the critical flaw population.

Strength test procedures can be grouped into various categories, e.g. flexural tests, direct tensile tests, compressive tests, and others.

The Fracture of Brittle Materials: Testing and Analysis, Second Edition.
Stephen W. Freiman and John J. Mecholsky, Jr.

FLEXURE TESTS

Flexure tests are the most straightforward, least expensive, and, therefore, the most popular means of measuring fracture strength. They are effective because most failures in ceramics begin from surface flaws,[1] Quinn (2016) which vary in size, shape, and orientation.

Flexure tests can be grouped into two categories, namely, uniaxial, which can be either three- or four-point bending, and biaxial. There are ASTM standards (ASTM C1161-13 2013) for both three-point flexure and four-point flexure tests as well as some of the biaxial flexure tests (ASTM C1499-15 2015).

Uniaxial Flexure Tests

A beam loaded in three-point flexure is illustrated in Figure 5.1. Either a rectangular or a circular beam can be used. The maximum stress occurs on a line at the center of the stressed length; for circular cross sections, however, the maximum stress occurs only at one point. The fact that the location of the peak stress is isolated means that the

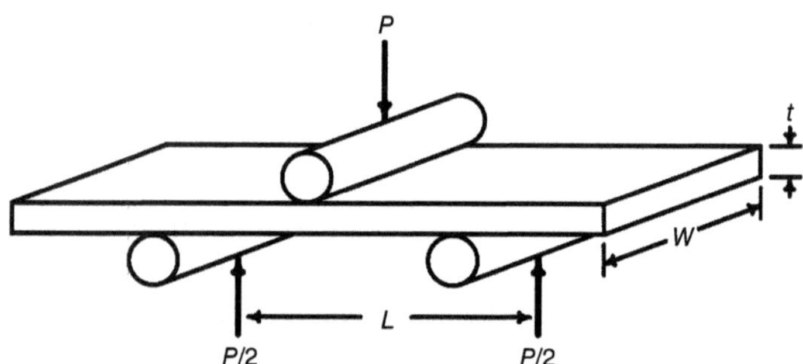

Figure 5.1 Schematic of three-point flexure test. P is the load, L is the span length, w is width, and t is thickness.

[1] By flaw we do not mean that there were necessarily errors in production, but that small surface cracks were created as a result of cutting and grinding. Nonetheless, if pores intersect the tensile surface of the flexure test, these can also act as sources of failure.

statistics of failure are not very good; the largest flaw may have been present at some distance from the maximum stress. One can compensate for the difference in location of the stress at the origin of failure, but this requires examination and measurement on each flexure bar. Details into the testing of rods are provided in Quinn (2016).

The maximum stress in a rectangular bar loaded at the center is along the tensile (bottom) surface (opposite the loaded side) at the center of the bar of thickness, t, and width, w, and is given by

$$\sigma = \frac{3PL}{2wt^2} \tag{5.1}$$

where P is the load and L is the distance between the outer supports. It is recommended for most ceramics and glasses (and not fiber composites) that the span to height ratio be between approximately 5 : 1 and 10 : 1 to avoid shear stress contributions for the shorter bars and large deflections affecting the calculations for the longer bars. Note that there is a possibility of wedging action at the loading point that may have to be accounted for. However, the corrections will be less than 8% for the worst case and for most cases will be approximately 4% (ASTM C1499-15 2015). By placing a compliant material under the loading points, the local loading stress can be distributed, and following ASTM C1161-13 (2013) guidelines, these wedging stresses can be minimized or eliminated.

If the origin of failure is not at the center opposite the loading point, then a correction for the stress a distance, x, away from the center point can be determined:

$$\sigma = \frac{3P(L-2x)}{2wt^2} \tag{5.2}$$

If the flexure bar fails at a distance away from the bottom of the specimen, i.e. not along the maximum tensile surface, then the correction for the failure a distance, y, into the depth is

$$\sigma = \frac{\left[(h-2y)/h\right]3PL}{2wt^2} \tag{5.3}$$

The deflection of the center of the beam, δ, can be determined:

$$\delta = \frac{PL^3}{48EI} \tag{5.4}$$

where E is the elastic modulus and I is the moment of inertia. Although three-point flexure can be used in the rapid evaluation of ceramics under development, four-point flexure testing is the more recommended method due to the larger area under stress, and therefore the greater sampling of the flaw population and better statistics.

A schematic of a four-point flexure beam is shown in Figure 5.2. The maximum stress in the central portion of the beam between the upper loading points on the opposite surface is constant:

$$\sigma = \frac{3PL_i}{wt^2} \tag{5.5}$$

where L_i is the distance between the outer support and the inner support. The most common arrangements for four-point flexure is quarter-point loading $L_i = L/4$ or third-point loading $L_i = L/3$. The possibility of wedging of the loading points is greater for four-point than for three-point flexure. Thus, the ability of the loading platens to be able to swivel is vital to successful tests. More details of these tests are described in the ASTM C1161-13 (2013) standard. The deflection of the maximum point, δ, is given by

$$\delta = \frac{PL_i\left(3L^2 - 4L_i^2\right)}{48EI} \tag{5.6}$$

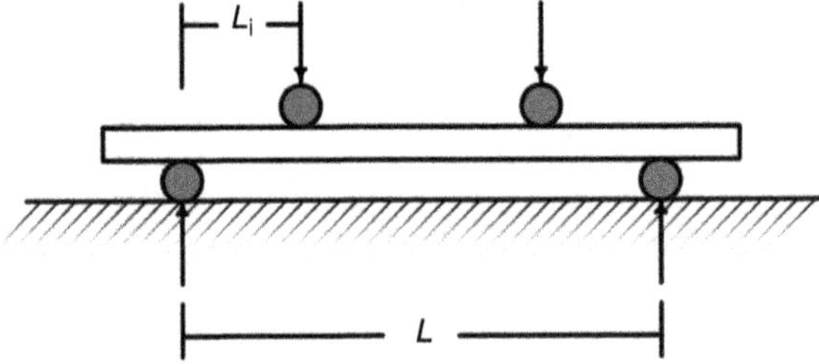

Figure 5.2 Schematic of four-point flexure test. Each arrow represents half the load, i.e. $P/2$.

Because of the increased area under uniform tension between the inner loading points, failure usually occurs on the tensile side between the loading points. However, fracture can occur into the depth away from the tensile surface. In these cases, a correction for the reduced stress needs to be made for the distance, y, into the depth:

$$\sigma = \frac{PL_{\mathrm{i}}}{wt^2\left(1-2y/t\right)} \tag{5.7}$$

Other Factors in Uniaxial Flexure Testing

Quinn and Morrell (1991) have nicely summarized the requirements and procedures to follow for flexure testing to be useful and meaningful, especially for design purposes. One of their conclusions is that fractography should be used whenever possible to understand the cause of fracture. It is recommended that an adhesive and compliant materials be placed under the loading point(s) to distribute the applied local stress due to the loading and to maintain the pieces after fracture. Mechanical machining or polishing cracks are often the sources of failure. Thus, if the surface is not carefully prepared, misleading information about the material strength may be obtained. In testing, the material should be loaded in a symmetric manner and evenly balanced. If it is not, there will be an uneven stress distribution, so that the strength at failure may not correspond to the load used in the calculation of the strength. Both the three- and four-point flexure procedures have been published in both ASTM and ISO standards to avoid these potential problems. Such standardization has been an important factor in reducing the cost of testing because of the reduced cost of manufacturing specimens all of the same dimensions.

In both types of uniaxial flexure tests, one must be concerned with the possibility of failure from damage at specimen edges; it is recommended that specimen edges be either rounded or chamfered to reduce or avoid this occurrence.

Failures in flexural testing can occur under the loading points due to incorrect alignment and inability to freely adjust the loading points.

Failure under the load points usually indicates an uneven loading configuration, and the data should be suspect. Likewise, failure outside the inner span should also be suspect and not used in any analysis.

BIAXIAL FLEXURE TESTS

If the edges of the final part will not experience significant stress during use, then biaxial flexure tests (Figure 5.3) are a better choice for measuring strength. The advantage of biaxial tests is that they eliminate edge effects and allows one to better distinguish effects of surface finish. In addition, the greater area under stress leads to better failure statistics. The five most common techniques are the ball-on-ring, ring-on-ring, piston-on-ring, piston-on-three-balls, and ball-on-three-balls (Figure 5.3a–d). Reviews of many of these have been written, e.g. Morrell et al. (1999) and Godfrey and John (1986). ASTM C1499-15 (2015) standard outlines the procedures for equibiaxial strength of advanced ceramics at ambient temperature via concentric ring configurations under monotonic uniaxial loading.

Ring-on-ring

In this test, a plate, which can be either circular or square, is supported on one ring while being loaded by a smaller ring (Ritter et al. 1980). A significant advantage of the test is the large area under constant stress (Figure 5.4). However, care must be taken to avoid wedging and frictional effects. In addition, stresses under the loading ring can be significantly larger than the average, leading to failure at these points. The solutions for the ring-on-ring test take the form

$$\sigma = \frac{AP}{t^2} \tag{5.8}$$

where σ is the maximum stress, P is the applied load on the ring, and t is the thickness. A is a constant that depends on the assumptions made for the analysis.

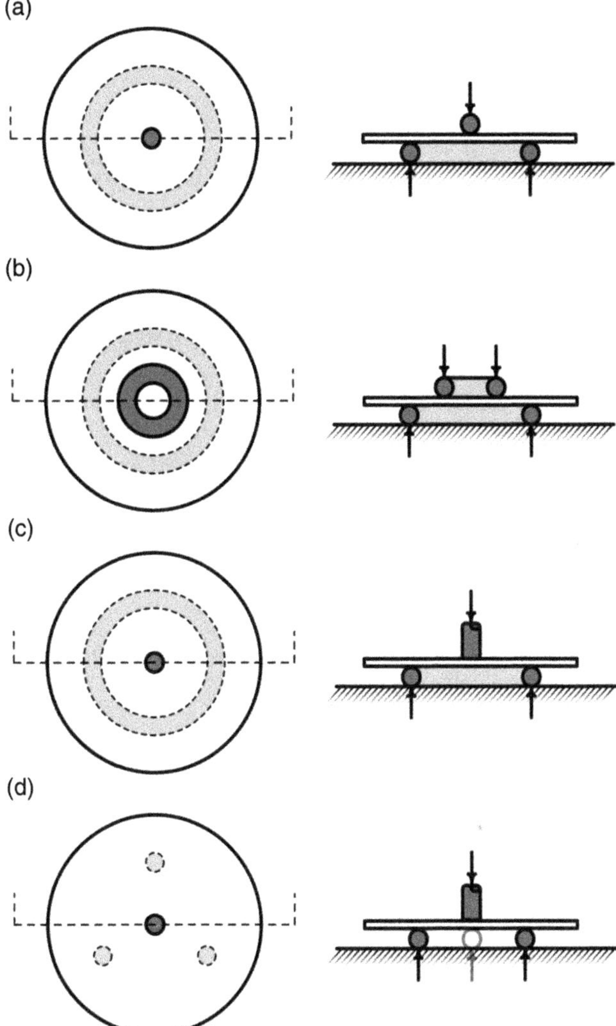

Figure 5.3 Schematic of different types of loading possible in biaxial flexure. (a) Ball-on-ring, (b) ring-on-ring, (c) piston-on-ring, and (d) piston-on-three-balls.

For example, one solution for the equibiaxial stress for the ring-on-ring test is given by Salem and Powers (2003):

$$\sigma = \left(\frac{3P}{2\pi t^2}\right)\left[\frac{(1-\nu)(D_S^2 - D_L^2)}{2D^2}\right]\left[(1+\nu)\left(\ln\frac{D_S}{D_L}\right)\right] \tag{5.9}$$

where P is the load, D is the plate diameter, D_L is the load ring diameter, D_S is the support ring diameter, σ is the equibiaxial stress, and ν is Poisson's ratio. The expression for the maximum displacement, δ, at the center is

$$\delta = \left[\frac{3P(1-\nu^2)D_L^{\,2}}{8\pi Et^3}\right]\left\{\left(\frac{D_S^2}{D_L^2}\right)\left[1+\frac{(1-\nu)(D_S^2 - D_L^2)}{2(1+\nu)D^2}\right]-\left[1+\ln\left(\frac{D_S}{D_L}\right)\right]\right\} \tag{5.10}$$

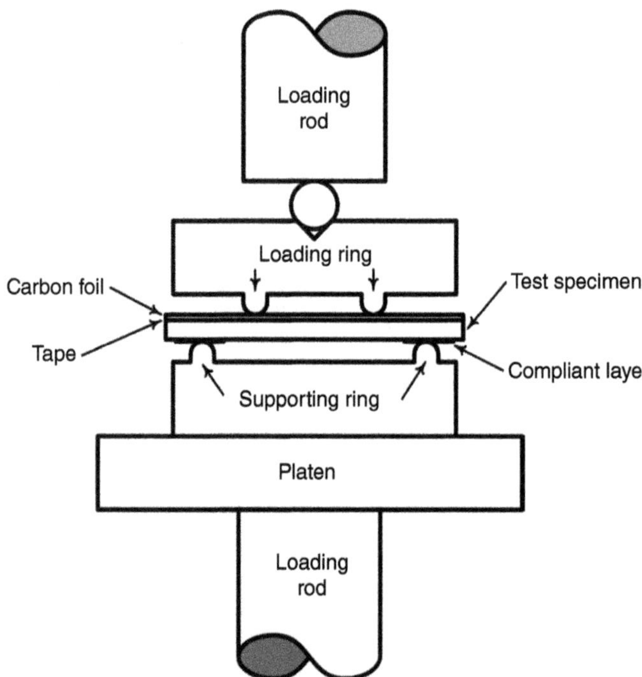

Figure 5.4 Schematic diagram of ring-on-ring loading.

where E is the modulus of elasticity and t is the thickness. Salem and Powers (2003) determined an estimate of the minimum plate thickness as

$$t > \left(\frac{2\sigma_f D_S^2}{3E} \right)^{1/2} \tag{5.11}$$

for which σ_f is the expected flexure strength.

Marshall (1980) modified the ring-on-ring test by introducing a liner spring for each ring. Each ring is supported on a rubber diaphragm covering a bed of fluid so that the system can be loaded hydraulically. The advantage is that equal bending forces are applied around the specimen/ring contact circle. If the loading is not coaxial, the system tilts the specimen. The disadvantage is that there is a nonlinear compliance. Thus, a constant displacement rate machine will not necessarily produce a constant loading rate. The basic equation is still Eq. (5.8), with a slight modification of "A." It is often important to validate the stress distribution of any loading configuration such as the ring-on-ring. This can be performed with numerical analysis or strain gages, e.g. cf. Hulm et al. (1998).

Ball-on-ring

In this test, rather than being loaded by a concentric ring, the specimen is supported on a ring, but loaded by a ball at its center. While in principle using a ball leads to a stress singularity, depending on the size and material of the ball, local deformation in the ball can alleviate this issue. The solutions for the radial, σ_r, and tangential, $\sigma\theta$, stresses on the tensile surface of a simply supported circular plate loaded over an area of small circular radius, r_0, can be obtained from any general mechanics text such as those of Roark and Young, *Formulas for Stress and Strain* (Young and Budynas 2001, http://www.roarksformulas.com). These standard equations were modified by Vitman and Pukh (1963) to account for the constraining effect of the annular overhang portion of the disk. The radial, σ_r, and circumferential, $\sigma\theta$, stresses are then

$$\sigma_r = \sigma_\theta = \sigma_{max} = \frac{3P(1+v)}{4\pi t^2} \left\{ 1 + 2\ln\left(\frac{a}{b}\right) + \frac{(1-v)}{(1+v)} \left[1 - \left(\frac{r_0'^2}{2a^2}\right)\left(\frac{a^2}{R^2}\right) \right] \right\} \quad \text{if } r < r_0 \tag{5.12}$$

$$\sigma_\theta = \frac{3P(1+v)}{4\pi t^2} \left\{ 2\ln\left(\frac{a}{r}\right) + \frac{(1-v)}{2(1+v)} \left[\frac{4-(r_0'^2/r^2)}{(a^2)(r_0'^2/r^2)(a^2/R^2)} \right] \right\} \quad \text{if } r > r_0 \tag{5.13}$$

where $r_0' = [1.6r_0^2 + t^2]^{1/2} - 0.675t$ for $r_0 < 0.5t$, $r_0' = r_0$ if $r_0 > 0.5t$, P is the load, ν is Poisson's ratio, a is the radius of the ring, t is the thickness of the disk, R is the radius of the disk, and r is the radial distance from the center of the disk. The value of r_0' is the equivalent radius of loading for concentrated loading. The value of r_0 is then the contact radius between the ball and the disk. Equations (5.8)–(5.10) are valid for maximum deflections less than half the thickness of the disk. Shetty et al. (1983) showed that by choosing a value of $r_0' = 0.33t$ ($r_0 = 0.1t$) for the ball-on-ring loading, generally good agreement was obtained between measured stresses using strain gages and those calculated using Eqs. (5.12) and (5.13). However, deWith and Wagemans (1989) suggest there is a more appropriate way of handling the overhang than that of Vitman and Pukh (1963):

$$\sigma_r = \sigma_\theta = \sigma_{max} = \frac{3P(1+v)}{4\pi t^2} \left\{ 1 + 2\ln\left(\frac{a}{b}\right) + \frac{(1-v)a^2}{(1+v)R^2} \left[1 - \left(\frac{b^2}{2a^2}\right) \right] \right\} \tag{5.14}$$

where b is the radius of uniform loading at the center. The appropriate selection of b is discussed thoroughly in deWith and Wagemans (1989).

Piston-on-three-balls

In the past, a very popular test was the piston-on-three-balls, in which a plate specimen is supported by three balls equally spaced about a circle and is loaded by a flat piston in the center of the circle. However, any deviation from parallel loading to the plane of the plate will lead to nonuniform stresses under the piston; thus, usually a mechanically absorbing material, such as Teflon, is used between the piston and the specimen. The fracture stress can be calculated from the fracture load using the following equation (Wachtman et al. 1972):

$$\sigma = -\frac{3}{4\pi}\frac{P}{t^2}(X - Y) \tag{5.15}$$

where $X = (1+\nu)\ln(b/R)^2 + [(1-\nu)/2](b/R)^2$, $Y = (1+\nu)[1+\ln(a/R)^2] + [(1-\nu)](a/R)^2$, P is the failure load, ν is Poisson's ratio, t is the specimen thickness, a is the support ring radius, b is the loading piston radius, and R is the specimen radius. There is much discussion of the use of this test technique. However, the test is still used in many industries because it minimizes the necessity of having completely flat and polished specimens. There is an ISO Standard 6872 (2015) that outlines the procedure for this test. If fractography and strain gages are used with the test, it can provide useful results.

A variation of this method is the piston-on-ring (Shetty et al. 1983; Hsueh et al. 2006). The stress distributions take the form shown in Eq. (5.8).

Ball-on-three-balls

In the ball-on-three-ball test (e.g. Danzer et al. 2006, 2007), a disk-shaped specimen is supported on three balls equidistant from its center. The opposite face is centrally loaded with a fourth ball normal to the orientation of the plane. The maximum principal tensile stress (σ_{max}) in the disk, located at the center on the surface opposite the loading ball, is the strength upon fracture of the disk. The stress field in the disk depends on the applied force and geometry (disk thickness and diameter, size, and the position of the balls) and also on the elastic properties of the ball and disk materials. The location of the fracture initiating crack can make a difference in the fracture pattern and strength (Jeong et al. 2002).

Different approaches have been made for the analytical calculation of the stress distribution in centrally loaded biaxial disk tests. They are all based on the linear elastic axisymmetric thin-plate theory. These approaches have been nicely summarized by Borger et al. (2002). Although the basic equation that defines the strength is Eq. (5.8), the actual stress distribution is usually determined using finite element analysis due to the many parameters that affect the results including the changing distribution of loading area as the load increases.

An outgrowth of this test that removes the nonuniform loading problem is to load the center of the plate with a ball containing a small flat on one side (Marshall 1980). The ball can be inserted into a socket to assure that it can rotate freely.

Other Factors

A question that frequently arises in each of these biaxial tests is the effect of specimen geometry, i.e. shape and degree of overhang of the support. Experience suggests that these factors have a minimal effect on the calculated stresses.

DIRECT TENSILE TESTS

Direct tensile tests are required if the flaws that can lead to failure are distributed throughout the volume of material. These flaws typically result from problems in processing, e.g. pores and inclusions. Several of the popular shapes used for tensile tests are shown in Figure 5.5. Due to their complex shapes, machining of these specimens is expensive. The two major issues associated with the tests themselves are gripping and alignment. Thus, most tensile testing procedures are focused on the adaptation of grips and the careful shaping of the specimens to avoid stress concentrations. While this book focuses primarily on room temperature tests, it should be noted that the gripping mechanisms become even more challenging at elevated temperatures. Alignment is a critical issue.

It is usually assumed that the tensile stress, σ, resulting from a load, P, distributed evenly across the original cross section, A, i.e.

$$\sigma = \frac{P}{A} \tag{5.16}$$

is uniformly applied along the length of the specimen in tensile testing. If there is a misalignment during loading, then a bending component, not accounted for in the analysis, is added to the stress, leading to errors in values obtained for the tensile strength. A tensile test in principle is simple. In actuality, it is one of the most difficult tests to perform properly because of alignment problems, gripping difficulties, and specimen preparation.

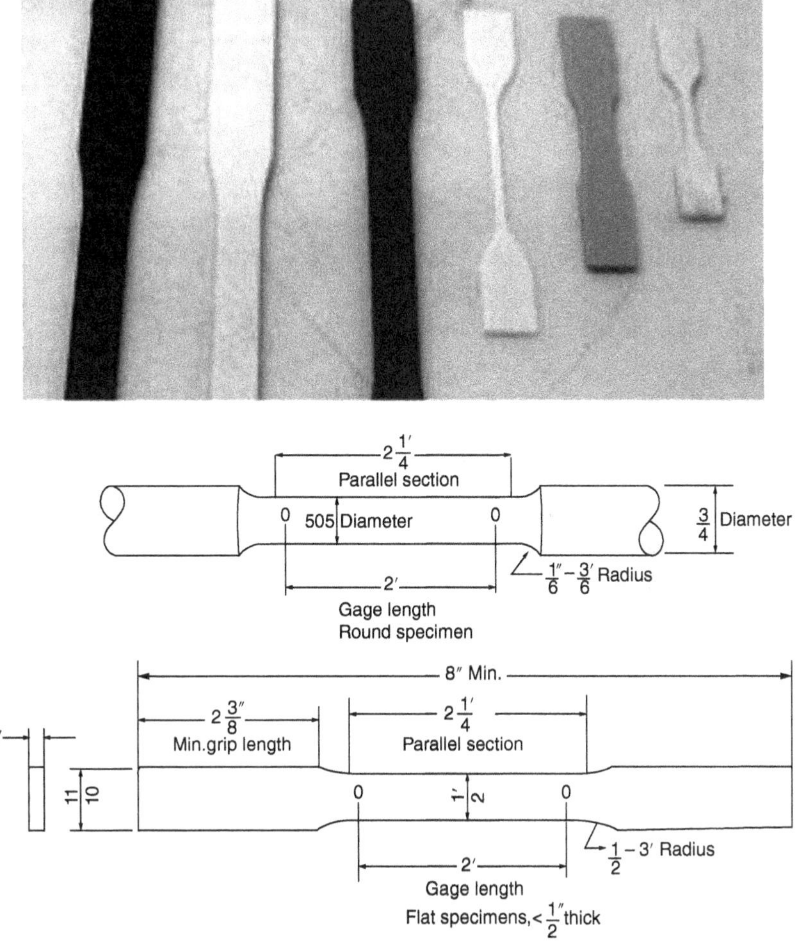

Figure 5.5 Tensile test specimen shapes. Both rectangular and circular cross sections are shown.

Uniaxial Tension

The main problem in direct tensile tests of brittle materials is the design of grips that will not damage the specimen. There are two separate categories of grip complexity: room temperature testing and high temperature testing. In room temperature testing, care must be made to make sure the loading points do not introduce stress concentrations so that failure occurs from the grips. The normal threads for metal specimens will not work with ceramics because of the stress concentrations of the threads and the negligible ductility and low toughness of the materials. Mostly dog-bone-shaped specimens are used in testing (Figure 5.5). These are either rectangular or cylindrical. The ends of these are flared out in order to reduce the stress near the contact points of the machine that will be used to pull the specimens in tension. There is a need for careful machining to get the proper radius to reduce the stress concentration over a large enough distance. These requirements lead to high costs in machining and material.

In elevated temperature testing, there are two types of grips that are considered: cold grips and hot grips. That is, either the grips are outside or inside the furnace. For cold grips outside the furnace, metals can be used far enough away so that the grips can be large to reduce stress. The disadvantage is that the material cost can be very high because the ceramic specimen now has to be relatively long. There is an effort to make small furnaces so that the amount of test material is reduced. In order to avoid thermal shock to the specimen, there needs to be a transition zone for the temperature gradient in the material. Otherwise there may be thermal shock during testing. High temperature alloys can be used to place the grips closer to the furnace, but this would require some cooling of the grips. Thus, cooling apparatus has been designed to help transition the temperature in the test material from cold to hot without thermal shock. These negative attributes have led to "hot grips" such that all the testing is done within the furnace. High temperature ceramic materials can be used as the grip material. Because of the brittleness of the grips, simple designs must be used. Because of the necessarily simpler designs, not the same detail can be applied to the alignment. Thus, when tensile testing at high temperature with the grips inside the furnace, care must be given to the design of the grips, specimen, and furnace. Often holes need to be machined into the test specimens. This machining requires very close tolerances and fine finish to avoid failure at the load points. All of these requirements increase the cost. Some examples of solutions to these issues are given in the following paragraphs.

Seshadri and Chia (1987) demonstrated simple self-aligning grips for tensile testing at room temperature (Figure 5.6). They showed that it is important to round the corners of the gripping blocks to minimize any stress concentration. This design produces reliable data, but the specimen needs to be relatively long in order to produce a generous radius for the grips and would increase the cost. In addition, Seshadri and Chia observed numerous failures outside the gage section.

Carroll et al. (1989) designed a high temperature fixture for measuring high temperature creep in tensile loading. Their method used hot grips and flat dog-bone specimens for testing up to 1500 °C. They used an extensometer to measure deformation by attaching "flags" on the specimens. Their technique required careful drilling of the holes in the specimen. It was important to place the holes along the centerline for proper alignment. They suggest using an electrical discharge machining (EDM) method for materials that are conducting because this reduces the cost compared with conventional grinding methods required for nonconducting materials.

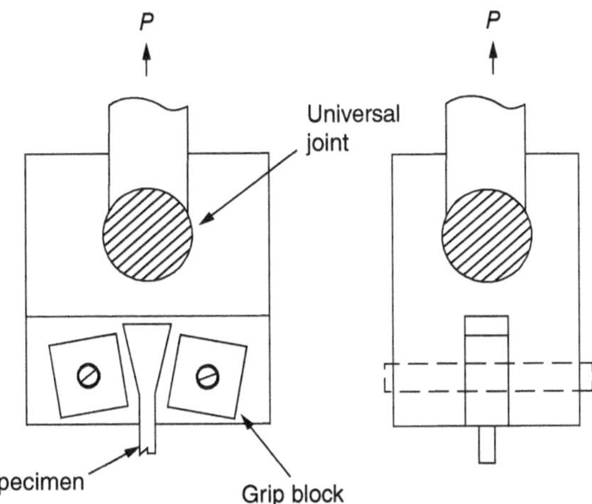

Figure 5.6 Schematic of self-aligning tensile grips. Source: Seshadri and Chia (1987). Reproduced with permission of John Wiley & Sons.

Klemensø et al. (2007) optimized the tensile specimen design for thin rectangular shapes. They used finite element analysis and experimental testing to reduce the stress concentration in the transition zone between the grips and the gage section. By optimizing the design, they were able to increase the number of specimens failing in the gage section. However, they found that most failures occurred due to the machining of the specimens. Because of their complexity and expense, direct tensile testing should only be conducted when it is clear that failures from non-surface flaws are important.

Fibers

Tensile testing of fibers is challenging because of the difficulty with gripping the ends of the fiber. Several techniques have been developed. Long-length fibers are spooled on a mandrel as the fiber comes out of the melt. Commercially, the fiber is tested while being processed from the melt by creating tension in one spool relative to another. This tension is created by having the take-up mandrel rotate faster than the supply mandrel. The tension essentially provides a proof stress for the fibers. However this test can yield misleading results for two reasons: (i) crack growth may occur upon loading and unloading, and (ii) fracture may occur before full tension is experienced in the fiber. In the case of stable crack growth occurring during loading, a methodology has been developed to account for the crack growth (Semjonov and Glaesemann 2007). This proof-testing methodology is discussed in Chapter 10. If unloading is not sufficiently rapid, crack growth can occur, weakening the fiber. The other problem is that fracture may occur on the mandrel before full tension is reached. In that case, one will assume that a greater stress has been achieved than actually occurs. This is not a safe condition. Fractographic analysis can be used to determine if the failure occurred with full load by examining if a symmetric mirror boundary exists or not (cf. Chapter 8).

For uniaxial tension tests of fibers in the laboratory, gripping usually involves clamps, serrated grips, and epoxy. Laboratory testing requires large number of tests so that meaningful statistics can be developed. However, controlled crack tests will provide fundamental information about fiber behavior using the intentional introduction of cracks via indentation (Dabbs and Lawn 1985).

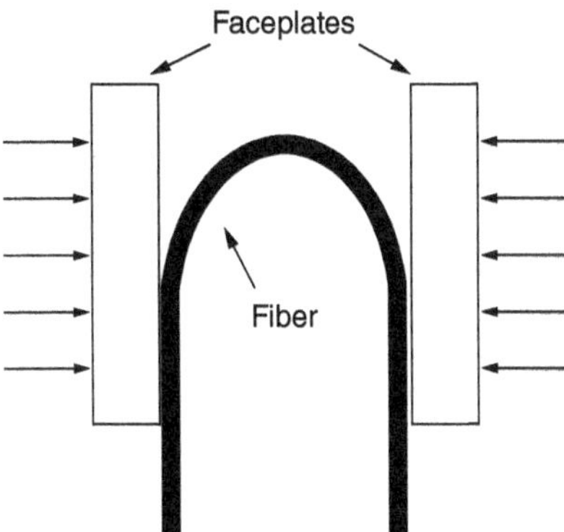

Figure 5.7 Schematic of two-point flexure apparatus.

A two-point bend test has been developed that obviates the gripping problem with tensile testing (Figure 5.7). This test is described in detail in several references (e.g. Matthewson et al. 1986; Matthewson and Kurkjian 1987). The basic concept is to bend a length of the fiber between two platens and measure the radius of curvature and distance between platens at fracture. The flexure strength, σ_{max}, can then be used to determine the potential tensile strength of the material:

$$\sigma_{max} = \frac{1.198E2r}{D-d} \tag{5.17}$$

where E is the elastic modulus, r is the fiber radius, D is the faceplate separation at fracture, and d is the overall fiber diameter (including coating).

OTHER LOADING CONFIGURATIONS

Rings in Flexure

Often, the components used in an application are hollow cylinders, such as those used in heat exchangers. It is often useful to be able to test the strength of these components in a similar fashion for which they are to be used. Thus, the "C-ring" and "O-ring" tests have been developed. These rings are usually prepared by machining sections from the cylinder. The testing procedure is outlined in ASTM

C1323-16 (2016) standard. The C-ring (Figures 5.8 and 5.9) and O-ring (Figure 5.10) can be loaded in compression or tension. Loading in this manner produces the maximum tensile stress in different locations, thus interrogating the distribution of flaws and surface finish on different surfaces. The O-ring loaded in compression (Figure 5.10) has two locations of maximum tensile stresses: (i) on the inside surface directly under the loading and (ii) on the outside surface 90° from the loading axis. Although these stresses are not of the same magnitude, (i) being the greater, the distribution of flaws, if different on the outside and inside surfaces, can make a difference as to the location of the failure.

The maximum tangential tensile stress, σ_f, on the inside surface in the plane of the loading for the O-ring of width, b, inside radius, r_i, and outer radius, r_o, loaded with P in compression, is calculated to be

$$\sigma_f = \frac{P}{\pi b r_o}\left[Q(r,\psi)\right] \tag{5.18}$$

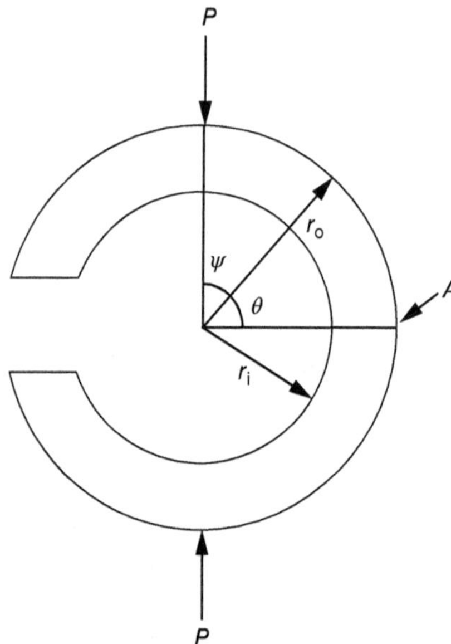

Figure 5.8 Schematic of C-ring specimen loaded in compression. Point A is the location of the maximum tensile stress.

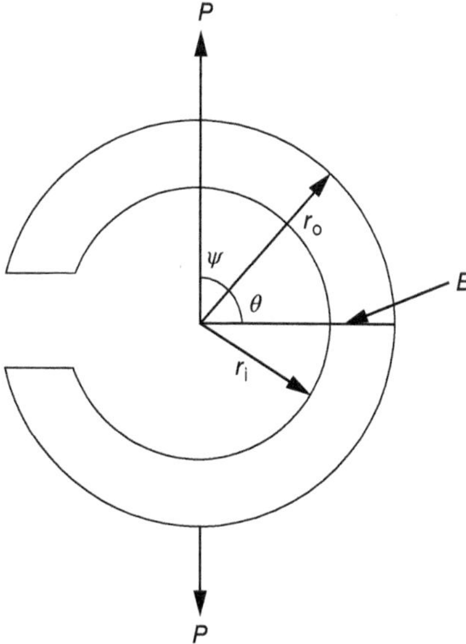

Figure 5.9 Schematic of C-ring loaded in tension. *B* is the location of the maximum tensile stress.

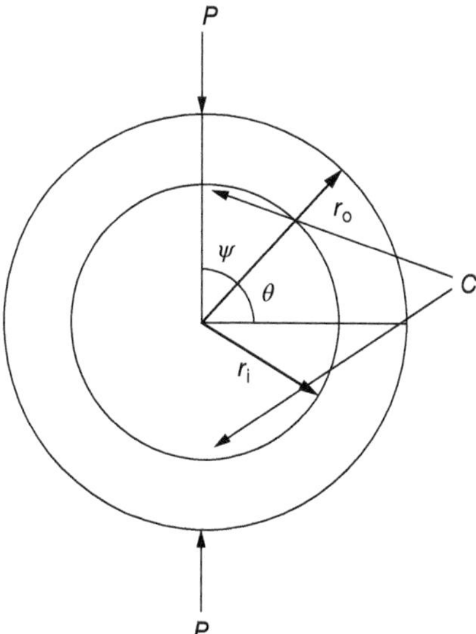

Figure 5.10 Schematic of O-ring loaded in compression. *C* indicates the location of the maximum tensile stress.

where

$$Q(r,\psi) = \left[\frac{1}{1-(r_i/r_o)^2}\right] - \left[\frac{(r_i/r_o)^2}{1-(r_i/r_o)^2}\right] + \sum_{n=2,4,\ldots}^{\infty} Z(n,r,\psi) \quad (5.19)$$

$Z(n,r,\psi)$ is a lengthy expression that accounts for the r_i/r_o ratio. However, if it is to be used, you should refer to the original references of Jadaan et al. (1991).

The O-ring loaded in tension has a similar stress distribution, except now the maximum tensile stress is located on the inside surface perpendicular to the load axis. It results in a similar equation as Eq. (5.18). The O-ring loaded in tension is not used as frequently as the one loaded in compression due to the difficulty of placing a loading point in tension in brittle materials.

The C-ring loaded in compression has the maximum tensile stress located on the outside surface on an axis perpendicular to the loading axis:

$$\sigma_f = \frac{P}{bt}\left[\frac{R(r_o - r_a)}{r_o(r_a - R)}\right] \quad (5.20)$$

where $r_a = (r_o + r_i)/2$ and $R = (r_o - r_i)/\ln(r_o/r_i)$, b = width, and t = thickness. The C-ring loaded in tension has a similar equation:

$$\sigma_f = \frac{P}{bt}\left[\frac{R(r_a - r_i)}{r_i(r_a - R)}\right] \quad (5.21)$$

A reexamination of the C-ring test by Jadaan and Wereszczak (2009) showed that $1 \le b/t \le 3$ (not 4 as recommended in ASTM C1323-16 (2016)) and $0.50 \le r_i/r_o \le 0.95$.

Circumferential Tension

It is often desirable to be able to determine the strength of a material in a form as close as possible to the shape for which it is to be used. For example, heat exchangers are tubular, and piezoelectric sonar transducers or ceramic seals are often circular rings. Hence, tests have been developed to test the strength of tubular or sections of cylindrical components (Ainscough and Messer 1974; Jadaan et al. 1991). The principle is relatively straightforward. A cylinder or ring is internally pressurized to create circumferential (hoop) tension in the section (Figure 5.11). Effecting the pressurization is difficult due to the fact that the pressurized medium must be contained. This complication leads to end conditions that must be considered in the stress analysis. Nevertheless, several techniques have been developed over the years to deal with the situation. Note that the O-ring specimen is loaded concentrically along one line whereas the specimens discussed here are loaded uniformly along the circumference.

For rings loaded by an internal pressure, p, the maximum tensile stress, $\sigma\theta$, is on the inside surface and can be calculated (Jadaan et al. 1991):

$$\sigma_\theta = \left(\frac{r_o^2 + r_i^2}{r_o^2 - r_i^2}\right)p \tag{5.22}$$

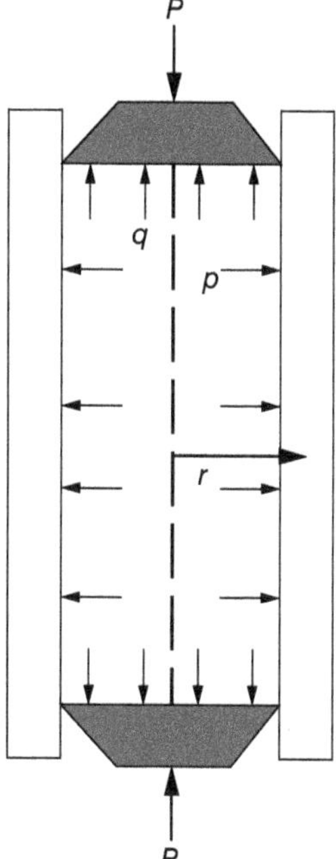

Figure 5.11 Schematic of the loading on a tubular component. Source: From Shelleman et al. (1991).

where r_i and r_o are the inner and outer radii, respectfully. At room temperature, the internal pressure, $p = p'$, can be applied using a rubber insert loaded axially, i.e. perpendicular to the ring diameter. The internal pressure, p', then can be calculated:

$$p' = \frac{\nu_r P}{(\pi r_i^2)(1-\nu_r)} \tag{5.23}$$

where ν_r is Poisson's ratio for rubber and P is the axial load. At high temperature, other means of applying the pressure must be designed. Some examples are provided in the literature, e.g. Ainscough and Messer (1974) and Jadaan et al. (1991). The stresses are still calculated using Eq. (5.22).

For tubular components, the stress calculation is again the same as for internally pressurized rings. Some modification needs to be accounted for at the ends depending on loading conditions. The details for loading an internally pressurized tubular component both at room temperature and high temperature are provided in Jadaan et al. (1992).

COMPRESSIVE LOADING

Diametral Compression Test

The diametral compression test, also known as the Brazilian disk test (Carniero and Barcellos 1953), was developed for materials that are easily fabricated in cylindrical form, e.g. concrete. The test involves loading a disk in compression along its diameter as shown in Figure 5.12, producing a tensile stress at the center of the disk. The advantage of this test is its simple specimen geometry, minimal requirements for fixtures, and its potential for application at elevated temperatures.

A biaxial stress state is generated in a specimen under diametral compression, with a compressive stress in the direction of loading and a tensile stress in the direction perpendicular to that of the loading. At the center of the disk, the magnitude of the tensile stress, σx, is given by

$$\sigma_x = \frac{2P}{\pi D t} \tag{5.24}$$

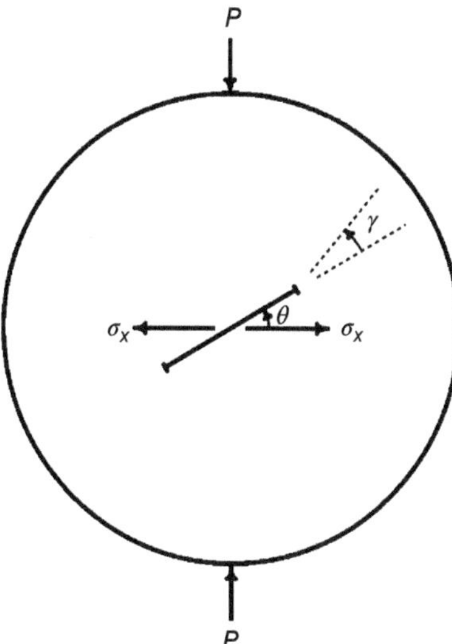

Figure 5.12 Schematic of the diametral compression test. The maximum tensile stress is at the center of the specimen. This specimen can be used to determine the effect of mixed-mode loading. When the crack is perpendicular to the loading, it is in pure mode I loading. This specimen is also called the Brazilian Disk Specimen.

where P is the load, D is the diameter of the disk, and t is the thickness. This value of stress remains approximately the same for a significant region near the center of the disk. The angle θ is with respect to the direction of the tensile stress. It is extremely useful to insert a piece of soft material, e.g. Teflon, between the loading platen and the specimen to avoid very high local stress concentrations and potential crushing.

Theta Specimen

The theta specimen is a hollow ring with a horizontal strip or web connecting two opposite sides of the ring (Figure 5.13). By applying a compressive load to the top of the ring, a tensile stress builds in the web as the specimen tries to expand. The ability to apply a compressive load to obtain a tensile stress alleviates the complications of alignment and gripping failures, especially for very small specimens. The advantage of this test is that it can be used for small components such as MEMS

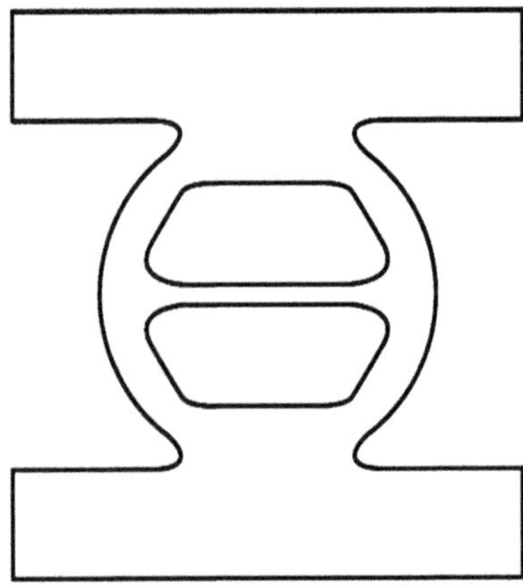

Figure 5.13 Schematic of theta specimen. Pure tension is generated in the center cross member when the specimen is loaded in compression.

devices. The original shape, created by Durelli et al. (1962) for bulk tests of brittle materials, has been optimized to prevent buckling in the ring and filleted to reduce secondary stress intensifications (Quinn et al. 2005). The tensile stress in the web is given by

$$\sigma = K\frac{P}{Dt} \tag{5.25}$$

where *D* is the diameter of the ring, *t* is the thickness of the sample, *P* is the applied force, and *K* is a constant that depends on the ratio of *w* to *D*. The original specimen had a diameter of 76 mm and a web thickness of 2.5 mm. Quinn scaled the same geometry to 250th of the size, creating specimen with a diameter of 300 µm and a web thickness of 7 µm (designed for 10 µm, but the final dimension was 7 µm) (Quinn et al. 2005). The relation for *K* was determined from finite element analysis to be approximately

$$K = 25.0 - 336\frac{w}{D} \tag{5.26}$$

This assumes the material is homogeneous and isotropic (Quinn et al. 2005). Finite element analysis was used to explore three possible methods that the indenter could contact the top of sample – surface contact, line contact, and point contact. In order to ensure that the specimens fail in the web, the stress in the web has to be substantively greater than any other secondary stress. This can be quantified by the ratio of the secondary stress to the web stress:

$$R_s = \frac{\sigma_s}{\sigma_w}$$

In the preferred but experimentally rare case of complete surface contact, $R_s = 0.75$. This ratio increases for line contact to 1.0 and is even greater for point contact at 1.2. In all cases, the maximum secondary stress occurred on the underside of the loading surface. By modifying the geometry of the sample to include a tab on the surface, the ratio for point contact drops to 0.54 (Fuller et al. 2007). If the axis of loading is not perpendicular to the top of the specimen, a stress gradient develops in the gage. This gradient is greatest in the line contact case, manageable with surface contact, and negligible for point contact.

Compressive Strength

Metallic materials are tested in compression to determine parameters such as ductile behavior. However, materials that fail in a brittle manner do not meet the requirements for axial compressive testing (ASTM C1424-15 2015). While the compressive strength of ceramics has been reported, compressive loading does not mean that those materials actually failed in compression. If there is a pore, crack, or other geometric defect in a brittle material and it is loaded in compression, there will be a tensile component of the stress around the defect. Thus, even though brittle materials fail while loaded in compression, failure actually results from the creation of a tensile stress. This fact can explain the large scatter in "compressive strength." There are other complications associated with compression testing of ceramics and other brittle materials such as cements, concrete, etc. We do not recommend performing compressive testing because it will provide misleading data. "Compressive strengths" should not be used for design or for calculations of reliability. This caution applies to the testing of biomaterials as well.

BIOMATERIALS

In the case of ceramic biomaterials such as zirconia and alumina used as replacements for bone, the tests mentioned previously in this chapter can be used. However, for some ceramics used for dental applications, other tests have been developed (e.g., Chai et al. 2011; Chai 2014). One of the most popular strength tests for dental biomaterials is the microtensile test (Pashley et al. 1999). The test is essentially the usual tensile test that is reduced in size to accommodate various systems. The microtensile strength is calculated as the maximum applied load at failure over the cross-sectional area. Several studies (Soares et al. 2008; Sadek et al. 2010) have described the difficulties and solutions to producing reliable microtensile tests. There are several different shape configurations suggested for the microtensile specimen (Sadek et al. 2010). These are shown in Figure 5.14. There are essentially three primary shapes, with slight variations from these shapes, for the microtensile tests: stick, dumbbell, and hourglass. Several studies have performed finite element analysis to determine the stress distributions in the various configurations (Soares et al. 2008; Sadek et al. 2010). In addition, attention must be paid to the cutting of the samples and the finish on the surfaces, as

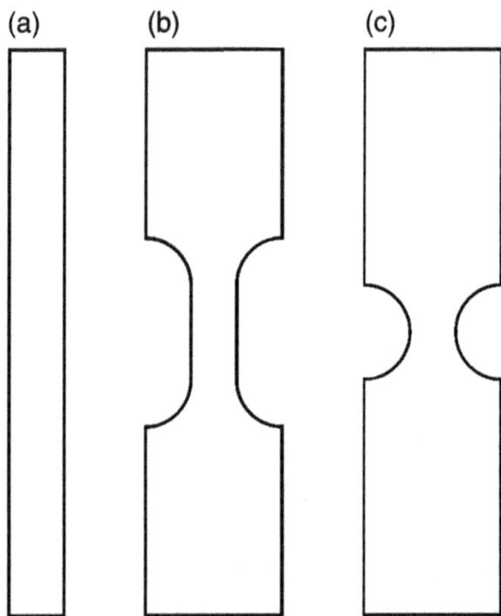

Figure 5.14 Microtensile configurations: (a) stick, (b) dumbbell, and (c) hourglass.

with all strength testing specimens. From a stress concentration viewpoint, the stick configuration offers the most straightforward approach to testing. If there was not an effect of the manufacturing of the sample, then the stick and dumbbell shape should provide the same strength values, everything else being equal. The stress concentration in the hourglass configuration is much different than the other two configurations and would, most likely, result in a lower strength. It is more difficult to produce the dumbbell and hourglass shapes during manufacturing, and the possibility of producing flaws in the process is greater in the dumbbell and hourglass configurations than the stick configuration.

Microtensile specimens are often used to test the adhesion of synthetic materials, such as ceramic or resin composites, to either dentin, the hard tissue situated beneath the enamel, or the enamel itself. The two materials are usually joined with an adhesive. The alignment of the two halves is even more critical than for homogenous specimens. If adhesion is being tested, the thickness of the adhesive interface is of prime importance. The modulus of the adhesive can affect the distribution of the differences in the maximum stresses in the two components. It is advised to use some analytical assessment such as finite element analysis along with the experimental testing when using the microtensile configuration for evaluating adhesion. It is also advised that the fracture surface be examined in adhesive tests to determine if the failure originated in the adhesive or at the adhesive–material interface, i.e. whether the failure was adhesive or cohesive.

Flexural Strength Testing of Laminates

Traditional laminate theory can be used to evaluate the strength of biomaterials that are laminates or can be modeled as laminates. If these specimens are beams, the theory has been well developed and exists in numerous textbooks on laminate theory (Timoshenko and Woinowsky-Krieger 1959). For laminates that are tested in biaxial flexure, a modification is needed.

The challenge is to convert experimental failure loads measured through biaxial flexure tests into stresses for laminated disks. A mathematical solution can be developed by combining laminate theory derived for the bending of beams with bending moments derived from simply supported, center-loaded circular plates (Timoshenko and Woinowsky-Krieger 1959; Gibson 2012). The failure stress, σ_f, for the laminates was determined using Eq. (5.27):

$$\sigma_f = P * \sigma'_{LT} \tag{5.27}$$

where P is the failure load (in units of N) measured experimentally and σ'_{LT} is the maximum principal stress (in MPa N^{-1}) calculated using laminate theory (Gibson 2012). The laminate theory solution calculates both the stresses and strains for the laminates, and therefore stress–strain curves are generated from the applied loads measured during biaxial flexure tests. Laminate theory can be used to describe the stress developed in flexure for fiber-reinforced cross-ply laminate beams (Gibson 2012). However, since the experimental design calls for biaxial flexure of laminated disks, not beams, the bending moments used in the laminate theory are modified to reflect the change in loading geometry. Descriptions are applied to the desired solution for bending moments of simply supported circular plates loaded at the center as described by Timoshenko and Woinowsky-Krieger (1959).

In summary, the calculation of stresses in a laminate plate loaded in biaxial flexure can be determined by modifying standard laminate theory to account for bending in a circular plate. A detailed procedure is explained in the Appendix at the end of this chapter.

An example of laminate analysis is shown in a model system to demonstrate a combination of analysis techniques in analyzing the fracture of multilayered systems. A laminate composite consisting of hydroxyapatite (HA) and polysulfone (PSu) layers was fabricated by tape casting the HA and stacking these layers with PSu interlayers using a solvent casting technique (Robinson et al. 2014). Laminate theory was applied to determine the strength of the laminates using bending moments derived from simply supported, center-loaded circular plates. Equation (5.27) was used to calculate the failure stress, σ_f, for the laminates.

Two methods were used to calculate stress in the middle HA layers. First, an experimental method of quantitative fractography was used to measure flaw sizes (Table 5.1). Failure loads for outer and middle layers were distinguished in non-indented laminates from measurable drops in the load–displacement curves or by extrapolating initial slopes. The

Table 5.1

Comparison of Failure Stresses Calculated from Fracture Mechanics and Laminate Theory

Laminate structure	n	Experimental failure load (N)	Middle layer flaw size (μm)	σ_f, flaw size calculation (MPa)	σ_f, laminate theory calculation (MPa)
400-200-800	13	155±22	254±41	30±2	28±6
100-200-800	1	77	314	27	44
200-200-800	3	112±18	178±30	36±3	39±6
100-200-1400	2	131±28	190±4	35±0	47±11
100-100-1600	6	107±19	289±81	29±4	35±6

peak load indicates failure of the middle HA lamina, which suggests that the flaw size in the middle layer controls the ultimate strength of the laminate. A constant fracture toughness of 0.6 MPa·m$^{1/2}$ was assumed for monolithic HA. Second, a theoretical method employed laminate theory in conjunction with the failure loads from load–displacement curves to calculate stresses in each layer using Eq. (5.27).

The positive feature of laminate design is that propagating cracks through the laminate thickness are impeded at the interfaces between HA and PSu lamina. This imposition results in either crack arrest for a given load or crack deflection along the interface. Fractographic evidence of crack origins in the middle HA later suggests that crack arrest and reinitiation occur in laminates. This reinitiation process contributes to the toughness observed in laminates in Figure 5.15, as compared with monolithic HA. Cracks are reinitiated in the middle layer and propagated through the thickness of the laminate in Figure 5.16. This reinitiation process contributes to the overall apparent toughness.

Note that if conventional beam theory is used in this analysis, the outer layer stresses would be approximately the same for thick outer lamina. However, for thin outer layers, the outer layer stresses would be overestimated. In addition, the calculations for the middle layer stresses, which control the failure process, would also be overestimated. Thus, to truly understand the distribution of stresses and their influence on failure mechanisms in biological systems that are laminated, laminate theory should be used to calculate the stresses.

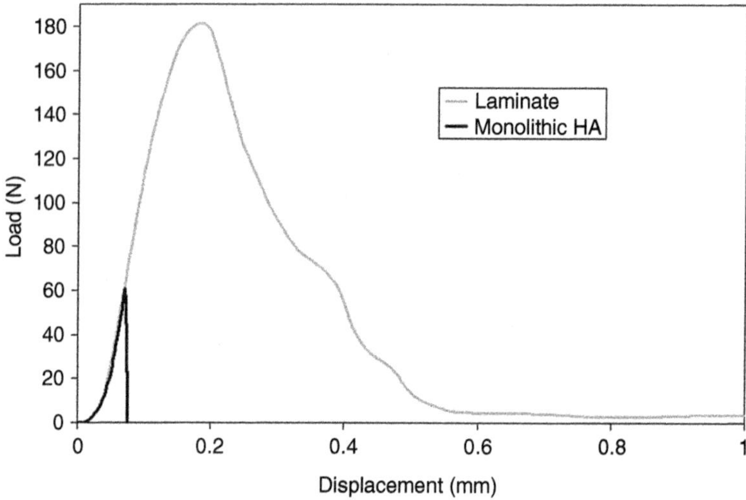

Figure 5.15 Comparison of load–displacement curves for the 400-200-800 laminates compared with monolithic HA.

Figure 5.16 SEM image of failure origin located in the middle HA layer. Arrows outline the reinitiated crack in the second HA layer.

STATISTICAL ANALYSIS OF STRENGTH DATA

The statistical nature of fracture in brittle materials, e.g. ceramics, semi-conductors, or brittle metal alloys, complicates the job of the engineer in designing and manufacturing parts that will not fail in service.

All brittle materials contain a statistical distribution of "flaws" that can lead to failure. As noted previously, we do not mean that there were defects in production, but that small surface cracks were created as a natural result of cutting and grinding. These cracks vary in size, shape, and orientation. The stresses on this flaw population that could lead to failure can arise, not just due to externally applied mechanical loads but also because of processing-induced stresses, thermal gradients, phase transformations, or the presence of applied electric fields.

There are several methods for comparing statistical distributions (Boslaugh and Watters 2008; Spiegel et al. 2009). The distribution that is most often applied to the strength of brittle materials is that of Weibull (1939). The Weibull distribution is based on the so-called "weakest link" approach that accounts for a volume (or area) dependence of the strength. This model assumes that failure from the most severe flaw leads to the failure of the entire structure. Others (e.g. Batdorf and Crose 1974; Lu et al. 2002) have shown that the Weibull distribution falls short in explaining failure in which mixed-mode fracture can occur. Nonetheless, it is a plausible model for most descriptions of strength distributions for ceramics.

In the Weibull model, the probability of failure, P_V, for a body of volume, V, can be generally expressed as

$$P(V)=1-\exp\left[-\int_V\left(\frac{\sigma-\sigma_u}{\sigma_\theta}\right)^m dV\right] \qquad \text{for}\ \sigma>\sigma_u \tag{5.28}$$

where σ is the applied stress; σ_θ is a normalizing parameter known as the characteristic strength, which corresponds to a probability of failure of 0.632; σ_u is the stress level below which no failures are expected; and m is the Weibull modulus, a unitless parameter describing the breadth of the strength distribution. Because of the difficulty in assigning a value to σ_u, it is frequently assumed that the minimum strength is zero, leading to what is termed a two-parameter Weibull plot. However, the two-parameter Weibull expression is unduly conservative since the likelihood of a failure at zero stress is vanishingly small. Modern techniques have been developed, which allow for the calculation of σ_u and its uncertainty (Fong et al. 2019).

For a two-parameter Weibull distribution, the determination of the two parameters, $\sigma\theta$ and m, can be obtained by taking the natural logarithm twice of Eq. (5.28) with $\sigma_u = 0$ and for tensile loading. This results in

$$\ln\left(\ln\frac{1}{P}\right) = \ln V - m\ln\sigma_\theta + m\ln\sigma. \tag{5.29}$$

The parameter m is the negative slope of a straight line for $\ln\ln(1/P)$ vs. $\ln\sigma$. The value for $\sigma\theta$ can be obtained from the intercept axis at $\ln\ln(1/P) = 0$. The survival probability is estimated in several ways. One example is to assume

$$P_i = 1 - \left(\frac{i-0.5}{N}\right) \tag{5.30}$$

or

$$P_i = 1 - \left(\frac{i}{N+1}\right) \tag{5.31}$$

where Pi is the probability of the ith specimen and N is the total number of specimens. Typically, the strengths of a set of specimens are ordered from least to greatest. Then the probability of fracture for each specimen is determined. The Weibull graph can be obtained from Eq. (5.29). Values of the Weibull parameters for use in the Weibull graph can be obtained in several ways, but the most common is through a maximum likelihood method (MLE) (Bury 1975). ASTM Standard C1239-07 or ISO Standard 6872 (2015) can be used as a guide in obtaining the Weibull parameters. The recommended number of specimens to test to obtain optimum results is 30. This is the point of diminishing returns in terms of the efficiency of testing (Lau 2017).

Examples of the effect of specimen size and shape are shown in Figures 5.17 and 5.18. In Figure 5.17, siliconized SiC parts were tested using several different configurations that included flexure bars, C-rings loaded in tension and compression, O-rings loaded in compression, and short and long tubular components. All specimens were taken from the same lot of material. The distribution of strengths shown in Figure 5.17 is to be expected based on the volume of material under stress in each. Those specimens having the smallest stressed volume (or area) show

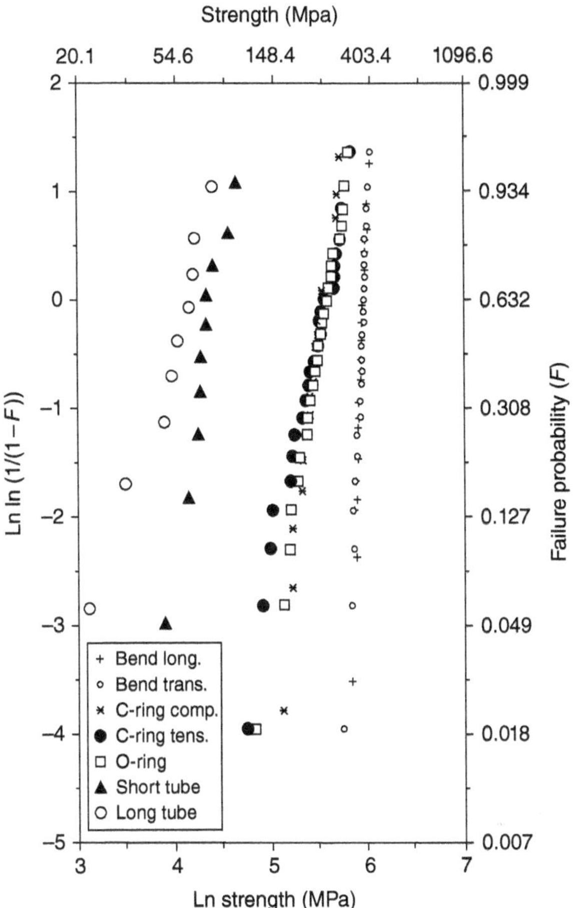

Figure 5.17 Weibull graph of strength distributions from various loading configurations on SiC specimens.

the greatest strength. No anisotropy in strength due to fabrication can be observed. More detailed information can come if this is combined with fractographic analysis as seen in Figure 5.18.

In Figure 5.18, the failure sources of the O-ring specimens loaded in compression and the C-ring specimens loaded in tension were identified. We can see that the lower strengths were caused entirely by silicon nodules on the tensile surface of the tubes. In the processing of these tubes, silicon is placed in the silicon carbide tubes, and the

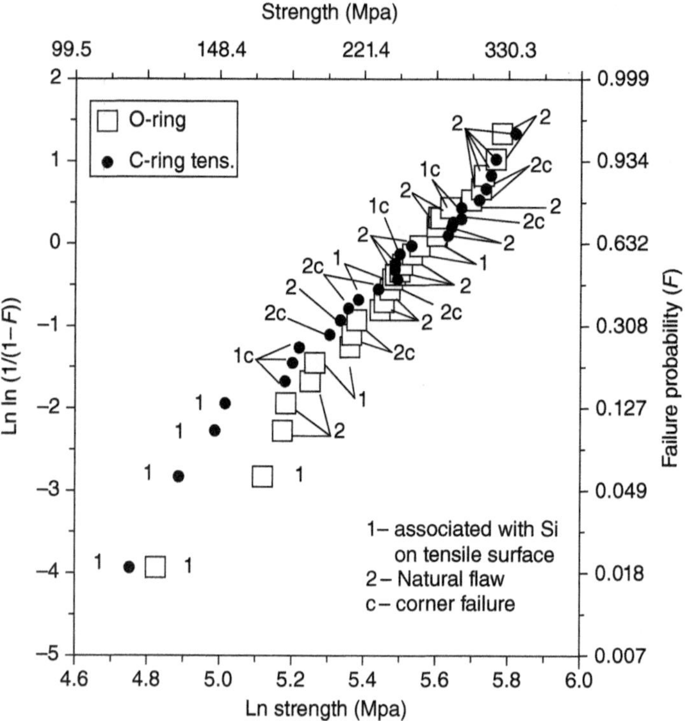

Figure 5.18 Strength distributions for SiC C-ring specimens loaded in tension and O-ring specimens loaded in compression. Notice that all the failure origins of the specimens have been identified. This type of analysis helps to understand processing, machining, or other sources of failure and helps obviate these.

silicon infiltrates the porous silicon carbide material during the heating process. Some residue is left on the inside surfaces of the tubes, while the outside surfaces are sandblasted "clean." The residual silicon also is suspected to result in residual stress. Thus, failure from the inside surface, when in tension, resulted in lower strengths than failure from the cleaned outside surfaces. It is strongly recommended that fracture origins be examined for specimens in fracture tests, whenever possible, particularly when the Weibull graph suggests that more than one type of flaw is contributing to failure. Details of the procedure for these forensic analyses are provided in the chapter on quantitative fractography.

Suggested Measurement and Analysis Procedures

The most important parameter in any prediction of survivability is the minimum stress, which could lead to the fracture of a part in service and knowledge of the uncertainty in this strength value. To ensure survivability over time, one needs a method to guarantee that the most severe flaw in a component is not subjected to such a stress.

One must use the statistical fracture behavior of the material as the measure of flaw severity. It is crucial to calculate not just the minimum stress at which failure could occur but also to know to what degree of confidence it can be specified. Since only a relatively small subset of the entire population of manufactured components is tested, it is important to have methods by which one can predict the possibility of failure in the components themselves.

Two-parameter Weibull statistics is at present the primary method by which strength distributions are fit and analyzed and is the method used in the ASTM standard (ASTM C1239-13 2018) to describe strength distributions for brittle materials (e.g., Medrano and Gillis 1987; Quinn 2003). The use of a two-parameter Weibull expression is not only unduly conservative and may not provide the best fit to most experimental strength distributions (Fong et al. 2019).

The average strength of a material may be useful for some applications, but, in most cases, knowledge of the minimum failure stress – and its uncertainty – is crucial. The most important experimental parameter that must be determined is the minimum possible strength in the distribution of manufactured parts. This minimum strength, S_m, is dependent on the flaw distribution in the as-produced material, a key assumption being that no flaws more severe than that leading to the minimum initial strength are introduced during service. It would be advantageous if the strength distribution of actual parts could be obtained under in-service loading conditions. However, this can be difficult and costly. Consequently, tests are usually conducted on small pieces of the same material. The processing procedures and surface treatments, e.g. machining and polishing, of these specimens must be identical to that of the component. Testing must be conducted at a high loading rate in an inert environment, e.g. dry gaseous nitrogen, to eliminate slow crack growth effects. Thirty specimens are typically recommended to assure that an adequate statistical distribution can be established. However, the actual number of specimens that one should test will depend on the scatter in the data and the standard deviation desired. If uniaxial flexural specimens are cut from a plate, one must make sure to randomize them with respect to the machining direction. Strengths perpendicular and parallel to the direction of grinding are different. If biaxial testing is employed, it is preferable to use a configuration putting the maximum possible area under stress, e.g. the ring-on-ring test, to obtain the most accurate statistical strength distribution. A component may experience a stress state that could cause it to fail from internal flaws such as pores or inclusions. In this case, flexural tests will not be effective, and direct tensile tests will be required.

In the simplest case, all flaws leading to failure are of the same type and come from one population, i.e. one source. In many instances, however, there are multiple flaw populations involved leading to more complex strength distributions. The presence of multiple flaw distributions usually manifests itself as a segmented strength-distribution plot. Fractographic analysis of the broken specimens, particularly those in the low-strength region, is important to determine the actual cause of failure and to separate unusually low strengths from those in the general population (Quinn 2016).

Methodology

We present a three-step approach to the calculation of the probability of failure of a given component.

Step 1: Determine a parameter estimation method, i.e. a function with which to fit the strength distribution of the set of specimens. The Weibull function, which is typically used in fracture analysis, is only one of many possible expressions that could be employed but has been used extensively because it is thought to represent the underlying physics governing brittle fracture. However, there is no evidence to support that, a priori, this function is superior to others.

The two-parameter Weibull distribution, which allows for failure at zero applied stress (Eq. 5.32), is used most often today, in part because of the complexities involved in calculating the parameters needed for a three-parameter Weibull distribution:

$$P_\mathrm{f} = 1 - \exp\left[-\left(\frac{S}{S_0}\right)^{m_2}\right] \tag{5.32}$$

The three-parameter Weibull distribution differs from the two-parameter function only in the introduction of a minimum lower strength, S_m:

$$P_\mathrm{f} = 1 - \exp\left[-\left(\frac{S - S_\mathrm{m}}{S_0}\right)^{m_3}\right] \tag{5.33}$$

where m_2 and m_3 each represent the shape of the curves but have different values.

In Fong et al. (2019) five expressions were compared as fits to data on a borosilicate glass (two-parameter Weibull, three-parameter Weibull, normal, two-parameter lognormal, and three-parameter lognormal). While there are large numbers of other possible functions that can be used to fit the data, the list is reduced by the requirement that the function must yield a positive *location parameter*, i.e. a minimum strength. The choice of a function to fit a data set should be based a goodness of fit as determined by one of several potential methods, e.g. the Anderson–Darling test (Anderson and Darling 1954), the Kolmogorov–Smirnov (KS) test, and the chi-square (CS) test (see, e.g. Bury 1975). For this demonstration we chose the KS criterion.

Step 2: Choose a method to fit the function to the data, i.e. an estimation method. The MLE (Bury 1975) is the most common method currently in use to fit strength distributions for brittle materials. This analysis will yield uncertainty bounds, e.g. 95% confidence limits, for the distribution of experimental data.

Step 3: The preceding steps refer to data taken on laboratory specimens; the strength data represents a sampling from the universe of strengths in the population of parts. Current procedure for a two-parameter Weibull distribution compares the stressed area (or volume) of a part to that of a specimen through the following expression:

$$\frac{S_{o2}}{S_{o1}} \propto \left(\frac{SE_1}{SE_2}\right)^{\frac{1}{m_2}} \tag{5.34}$$

where S_{o1} and S_{o2} are the strengths of the specimen and the component, respectively, and SE_1 and SE_2 are the corresponding effective stressed areas. However, Eq. (5.34) is applicable only to a two-parameter Weibull distribution.

A more modern approach is to employ the concepts of "tolerance limit" and "coverage" (Nelson et al. 2003), which yields the probability of failure for an entire population of parts rather than that of a single part.

We can explain these concepts as follows. The simplest confidence interval for the measurements taken from a series of test specimens describes the boundaries on the mean of the specimen population, $\bar{y} \pm d_1$. The lower confidence bound (LCB) is defined as

$$LCB = \bar{y} - d_1 \tag{5.35}$$

$$\text{where} \quad d_1 = t(0.025; n-1) * sd * n^{-\frac{1}{2}}. \tag{5.36}$$

t is the Student's t distribution, which is a function of the desired uncertainty (here 0.05) and the degrees of freedom $(n-1)$, sd is the standard deviation, and n is the number of specimens. For example, the 95% confidence limits describe the boundaries on the mean value 19 times out of 20.

There exists a second boundary, d_2, known as the one-sided lower prediction bound (LPB), defined as follows (Nelson et al. 2003):

$$LPB = \bar{y} - d_2 \tag{5.37}$$

$$\text{where} \quad d_2 = t(0.025; n-1)\, sd \left(1 + \frac{1}{n}\right)^{1/2} \tag{5.38}$$

This definition is a statement that a single future observation will be found to the right of the bound. Note that even when n approaches infinity, d_2 is never zero and is less than the sample mean.

Instead of working with a single future observation as in the definition of the LPB, we are interested in a certain proportion, p, of all future observations from the entire population of parts. The tolerance limit, d_3, makes a statement about the strength limits of a given proportion of the entire population of parts. We can calculate this limit to predict the minimum strength of a specific proportion of the set of the full-scale structure or components:

$$d_3 = r(n,p)u(n-1,\gamma)sd \tag{5.39}$$

$$LTL(p) = y - d_3 \tag{5.40}$$

For normal distributions γ and μ are numbers obtained from a statistical table (Nelson et al. 2003); their value depends on the degree of assurance desired. The position of each of these boundaries is shown in Figure 5.19 for a fictitious material with a normal distribution of strengths.

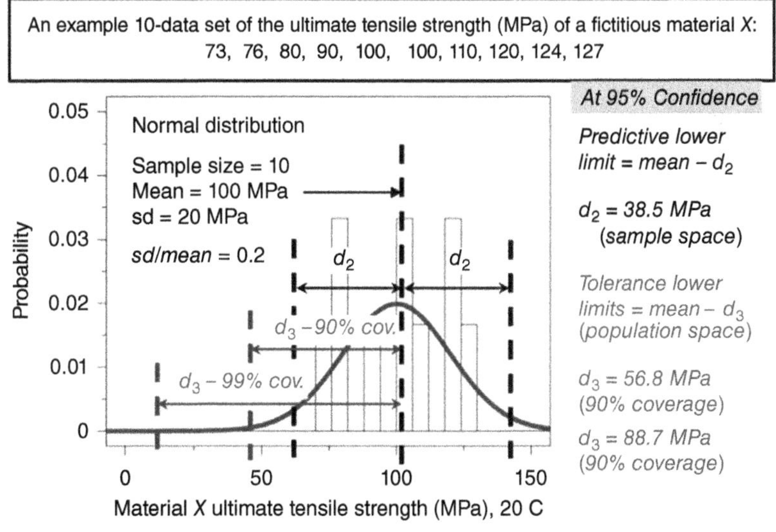

Figure 5.19 Positions of d_1, d_2, and d_3 boundaries for a normal distribution of strengths.

Note that prior to now, this procedure could only be used for normal distributions. A recent publication (Fong et al. 2019) demonstrates how to apply this procedure to other distributions, e.g. Weibull.

Coverage is defined as the percentage of the total number of parts (or in other words the total area of all parts under stress) to which we want the tolerance limit to apply. These uncertainty bounds for the lower limit to the strength will be larger than those for the specimen population alone.

SUMMARY

There are many testing configurations available to determine the strength of brittle materials (Table 5.2). Selection of a test should be based on the size and form of the final product, its end use, material costs, etc. The kind of test and the number of specimens tested will depend on many factors, not the least of which is the potential cost (in both dollars and injuries) of the failure of a given component. There are also sources available for providing the required level of detail in testing procedures, e.g. ASTM standards. We strongly suggest consulting the appropriate standard for the details to obtain accurate and reliable data.

We have also discussed the statistical analysis procedures that must be undertaken to make reliable predictions of the strength of actual components based upon the specimen test data.

Finally, we made the point that important information can be gleaned from the tests when the strength analyses are complemented by fractographic analysis of the fracture. More importantly, knowledge of the source of failure can be critical to decisions about test procedures and predictions of safety.

QUESTIONS

1. **(a)** In order to test the strength of a ceramic, cylindrical specimens of length 25 mm, of thickness 0.5 mm, and of diameter 5 mm are placed in axial tension. The tensile stress σ that causes 50% of the specimens to

Table 5.2
Summary of Advantages and Disadvantages of Useful Strength Tests

Test (reference)	Equation	Pros	Cons
Flexure tests: three- or four-point (ASTM C1161-13 2013; ASTM C1684-18 2018; ASTM C1211-18 2018; ASTM C1239-13 2018)	$\sigma = CPL/(bt^2)$ (rectangular beam) σ, strength; P, load to failure; L, span length; b, width; t, thickness; C, constant depending on loading. $C = 3/2$ for three-point flexure. $C = 1$ for 1/3 point four-point flexure. $C = 3/2$ for 1/4 point four-point flexure	Relatively simple to apply and calculate	Surface data only Four-point flexure needs to be aligned properly Three-point may not failure at center Width-to-span size is important
Uniaxial tension test (ASTM C1273-15 2010; ASTM C1239-13 2018)	$\sigma_T = P/A$ σ_T, strength; P, load; A, cross-sectional area	Surveys entire material	High cost of sample preparation Undesirable failure at loading points Potential of unknown flexural stresses
Diametral compression tests (ASTM C1323-16 2016)	$\sigma_x = 2P/\pi Dt$ σ_x, strength; P, load; D, diameter of disk; t, thickness of disk	Low cost of sample preparation Avoids edge conditions Able to evaluate mixed-mode loading	Requires method of distributing stress at loading points Should verify failure mode to validate test Potential buckling problem

(Continued)

Table 5.2
(Continued)

Test (reference)	Equation	Pros	Cons
Biaxial flexure (ASTM C1499-15 2015; Marshall 1980)	$\sigma = AP/t^2$ σ, strength; P, load; t, thickness; A, constant depending on specific loading configuration	Avoids edge conditions Able to test as fabricated platelike structures	Need to identify failure origin Stress can be concentrated at a load point
Theta specimen (Durelli et al. 1962)	$\sigma = K(P/Dt)$ σ, strength in web; P, load; t, thickness; D, diameter of ring; K, constant	Can develop pure tension from compressive loading	Difficult to prepare specimen Failure may not occur in web – need many specimens Need to distribute stress at loading points

	Standard
Flexural strength	C1161-13 – "Standard Test Method for Flexural Strength of Advanced Ceramics at Ambient Temperature" ISO 14704:2016 – "Fine Ceramics (Advanced Ceramics, Advanced Technical Ceramics) – Test Method for Flexural Strength of Monolithic Ceramics at Room Temperature"
Flexural strength at elevated temperatures	C1211-13 – "Standard Test Method for Flexural Strength of Advanced Ceramics at Elevated Temperature" ISO 17565:2003 – "Fine Ceramics (Advanced Ceramics, Advanced Technical Ceramics) – Test Method for Flexural Strength of Monolithic Ceramics at Elevated Temperature"
Tensile strength at room temperature	C1273-15 – "Standard Test Method for Tensile Strength of Monolithic Advanced Ceramics at Ambient Temperatures" ISO 15490:2008 – "Fine Ceramics (Advanced Ceramics, Advanced Technical Ceramics) – Test Method for Tensile Strength of Monolithic Ceramics at Room Temperature"

Table 5.2

(Continued)

	Standard
Tensile strength at elevated temperatures	C1366-04 – "Standard Test Method for Tensile Strength of Monolithic Advanced Ceramics at Elevated Temperatures"
Monotonic biaxial strength	ASTM C1499-15 – "Determination of Monotonic Biaxial Flexural Strength Advanced Ceramics"
Piston-on-3-balls	ISO 6872 standard
C-ring tests	ASTM C1323-16 – "Standard Test Method for Ultimate Strength of Advanced Ceramics with Diametrally Compressed C-Ring Specimens at Ambient Temperature"

fracture is 120 MPa. Cylindrical ceramic components of length 50 mm, of thickness 1 mm, and of diameter 10 mm are required to withstand an axial stress σ_c with a survival probability of 99%. Given that $m = 5$, determine σ_c.

(b) What would be the expected mean strength if the larger cylinders were loaded in four-point flexure? In other words, how would this new request change the calculations above?

2. You are developing a new polycrystalline ceramic material to be used as a substrate for a microelectronic device and want to evaluate the strength as you alter the composition. Which test would you use and why? How would you select the dimensions of the test?

3. **(a)** Assume you can test the same size specimen of an amorphous solid such as soda-lime-silica glass in uniaxial tension, three-point flexure, and four-point flexure. Would the test produce the same strengths? If not, which would be greater and which would be least? Please justify your answer with a rationale.

(b) Would the above answer be different for a sintered polycrystalline fine-grain ceramic? Why?

(c) Would the answer to (a) be different for a single crystal? Why?

4. You produce fine-grain polycrystalline ceramics in the form of solid cylinders that are 40 mm in diameter and 200 mm in length. They are to be used in disk form as a substrate. You want to determine the strength of the material. How would you proceed? [Hint: What test would you select? What size samples would you choose? Why? How many samples would you consider testing? What statistical and failure analyses would you perform?]

5. In performing microtensile tests of polycrystalline materials, how would you determine the appropriate dimensions of the specimen? Would there be a difference for materials with grain sizes in submicron versus materials with grain sizes of 100 μm? Why?

APPENDIX 5.A LAMINATE ANALYSIS EXAMPLE

Figure 5.A.1 shows a schematic of a laminate showing the variables necessary for the laminate theory calculations (Mallick 1993). First, a stiffness matrix Q is calculated in Eq. (5.A.1):

$$|Q| = \begin{vmatrix} Q_{11} & Q_{12} \\ Q_{21} & Q_{22} \end{vmatrix} \tag{5.A.1}$$

$$Q_{11} = Q_{22} = \frac{E}{1-\nu^2} \tag{5.A.2}$$

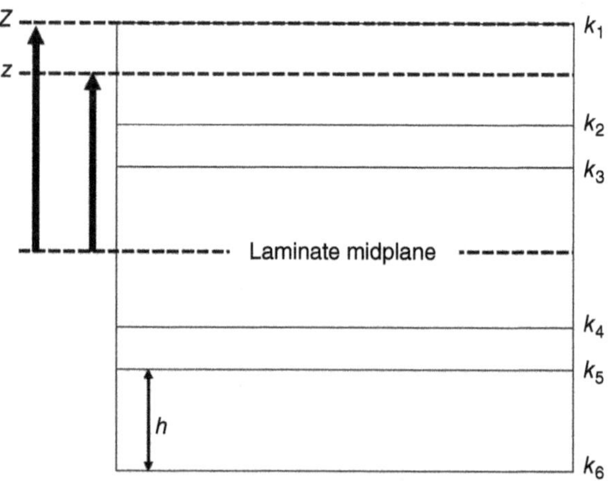

Figure 5.A.1 A schematic representation of a laminate indicating the variables required for calculations.

$$Q_{12} = Q_{21} = \frac{\nu E}{1-\nu^2} \tag{5.A.3}$$

where E is the elastic modulus and ν is Poisson's ratio of the particular lamina materials under analysis. From components of the stiffness matrix, the bending matrix D is calculated in Eq. (5.A.4):

$$|\boldsymbol{D}| = \begin{vmatrix} D_{11} & D_{12} \\ D_{21} & D_{22} \end{vmatrix} \tag{5.A.4}$$

$$D_{11} = D_{22} = \sum_{k=1}^{N} \left(\frac{Q_{11(k)} h_{(k)}^3}{12} + Q_{11(k)} h_{(k)} z_{(k)}^2 \right) \tag{5.A.5}$$

$$D_{12} = D_{21} = \sum_{k=1}^{N} \left(\frac{Q_{12(k)} h_{(k)}^3}{12} + Q_{12(k)} h_{(k)} z_{(k)}^2 \right) \tag{5.A.6}$$

For N layers, z is the distance from the midplane of the laminate to the centroid of the kth layer (see Figure 5.A.1), and h is the lamina thickness. An equivalent value for Poisson's ratio of the composite is calculated by Eq. (5.A.7):

$$\nu_{eq} = \frac{D_{12}}{D_{11}} \tag{5.A.7}$$

The next step in the solution is calculating the bending moments of inertia, M_r and M_t, through Eqs. (5.A.8) and (5.A.9), respectively. M_r corresponds to the circumferential sections of the circular plate, and M_t acts along the diametral sections of the plate:

$$M_r = \frac{P}{4\pi}\left(1+\nu_{eq}\right)\log\frac{a}{r} \tag{5.A.8}$$

$$M_t = \frac{P}{4\pi}\left[\left(1+\nu_{eq}\right)\log\left(\frac{a}{r}\right)+\left(1-\nu_{eq}\right)\right] \tag{5.A.9}$$

where P is the applied load, ν_{eq} is determined from laminate theory, a is the support radius, and r is the radial distance from the center of loading.

Using the inverse bending matrix and moments of inertia, laminate curvatures, K_r and K_t, are calculated in Eq. (5.A.10):

$$\begin{Bmatrix} K_r \\ K_t \end{Bmatrix} = \boldsymbol{D}^{-1} \begin{Bmatrix} M_r \\ M_t \end{Bmatrix} \quad (5.A.10)$$

The curvatures are used to calculate the strains ε_r and ε_t through matrix multiplication in Eq. (5.A.11):

$$\begin{Bmatrix} \varepsilon_r \\ \varepsilon_t \end{Bmatrix} = Z \begin{Bmatrix} K_r \\ K_t \end{Bmatrix} \quad (5.A.11)$$

where Z is the distance from the laminate midplane to the tensile surface of the kth layer and differs from the previous value for z.

Finally, the stiffness matrix Q and the strain values are used to calculate the stresses σ_r and σ_t in Eq. (5.A.12):

$$\begin{Bmatrix} \sigma_r \\ \sigma_t \end{Bmatrix} = \boldsymbol{Q} \begin{Bmatrix} \varepsilon_r \\ \varepsilon_t \end{Bmatrix} \quad (5.A.12)$$

In summary, the calculation of stresses in a laminate plate loaded in biaxial flexure can be determined by modifying standard laminate theory to account for bending in a circular plate.

REFERENCES

Ainscough, J.B. and Messer, P.F. (1974). An apparatus for the tensile testing of ceramic ring specimens at elevated temperatures. *J. Phys. E Sci. Instrum.* 7: 937–939.

Anderson, T.W. and Darling, D.A. (1954). A test of goodness of fit. *J. Am. Stat. Assoc.* 49: 765–769.

ASTM C1161-13 (2013). *Standard Test Method for Flexural Strength of Advanced Ceramics at Ambient Temperature*. West Conshohocken, PA: American Society for Testing and Materials.

ASTM C1211-18 (2018). *Standard Test Method for Flexural Strength of Advanced Ceramics at Elevated Temperatures*. West Conshohocken, PA: American Society for Testing and Materials.

ASTM C1239-13 (2018). *Standard Practice for Reporting Uniaxial Strength Data and Estimating Weibull Distribution Parameters for Advanced Ceramics*. West Conshohocken, PA: American Society for Testing and Materials.

ASTM C1273-15 (2010). *Standard Test Method for Tensile Strength of Monolithic Advanced Ceramics at Ambient Temperatures*, vol. 15.01. West Conshohocken, PA: American Society for Testing and Materials.

ASTM C1323-16 (2016). *Standard Test Method for Ultimate Strength of Advanced Ceramics with Diametrally Compressed C-Ring Specimens at Ambient Temperature*. West Conshohocken, PA: American Society for Testing and Materials.

ASTM C1424-15 (2015). *Standard Test Method for Monotonic Compressive Strength of Advanced Ceramics at Ambient Temperature*. West Conshohocken, PA: American Society for Testing and Materials.

ASTM C1499-15 (2015). *Standard Test Method for Monotonic Equibiaxial Flexural Strength of Advanced Ceramics at Ambient Temperature*, Annual Book of ASTM Standards, vol. 15.01, 779–788. West Conshohocken, PA: American Society for Testing and Materials.

ASTM C1684-18 (2018). *Standard Test Method for Flexural Strength of Advanced Ceramics at Ambient Temperature – Cylindrical Rod Strength*. West Conshohocken, PA: American Society for Testing and Materials.

Batdorf, S.B. and Crose, J.G. (1974). A statistical theory for the fracture of brittle structures subjected to nonuniform polyaxial stresses. *J. Appl. Mech.* 6: 459–464.

Borger, A., Peter, S., and Danzer, R. (2002). The ball on three balls test for strength testing of brittle discs: stress distribution in the disc. *J. Eur. Ceram. Soc.* 22: 1425–1436.

Boslaugh, S. and Watters, P.A. (2008). *Statistics in a Nutshell*. Sebastopol, CA: O'Reilly.

Bury, K.V. (1975). *Statistical Models in Applied Science*. New York: Wiley.

Carniero, F.L.L.B. and Barcellos, A. (1953). *Concrete Tensile Strength*, No. 13. Paris: Union of Testing and Research Laboratories for Materials and Structures.

Carroll, D.F., Wiederhorn, S.M., and Roberts, D.E. (1989). Technique for tensile creep testing of ceramics. *J. Am. Ceram. Soc.* 72 (9): 1610–1614.

Chai, H. (2014). On the mechanical properties of tooth enamel under spherical indentation. *Acta Biomater.* 10: 4852–4860.

Chai, H., Leeb, J.J.W., and Lawn, B.R. (2011). On the chipping and splitting of teeth. *J. Mech. Behav. Biomed. Mater.* 4: 315–321.

Dabbs, T.P. and Lawn, B.R. (1985). Strength and fatigue properties of optical glass fibers containing microindentation flaws. *J. Am. Ceram. Soc.* 68 (11): 563–569.

Danzer, R., Supancic, P., and Harrer, W. (2006). Biaxial strength tests for brittle ceramic rectangular plates. *J. Ceram. Soc. Jpn.* 114 (11): 1054–1060.

Danzer, R., Harrer, W., Supancic, P. et al. (2007). The ball on three balls test – strength and failure analysis of different materials. *J. Eur. Ceram. Soc.* 27: 1481–1485.

deWith, G. and Wagemans, H.H.M. (1989). Ball-on-ring test revisited. *J. Am. Ceram. Soc.* 72: 1538–1541.

Durelli, A.J., Morse, S., and Parks, V. (1962). The theta specimen for determining tensile strength of brittle materials. *Mater. Res. Stand.* 2: 114–117.

Fong, J.T., Filliben, J.J., Heckert, N.A. et al. (2019). Estimating with uncertainty quantification a minimum design allowable strength for a full-scale component or structure of engineering materials. *J. NIST Res.* (To be published).

Fuller, E.R. Jr., Quinn, G.D., and Cook, R.F. (2007). Strength and fracture measurements at the nano scale. *AIP Conf. Proc.* 931: 156.

Gibson, R.F. (2012). *Principles of Composite Material Mechanics*. Boca Raton, FL: CRC Press.

Godfrey, D.J. and John, S. (1986). Disc flexure tests for the evaluation of ceramic strength. In: *Proceedings of the 2nd International Conference of Ceramic Materials and Components for Engines*, 657–665. Lübeck-Travemünde: Verlag Deutsche Keramische Gesellschaft.

Hsueh, C.-H., Luttrell, C.R., and Becher, P.F. (2006). Analyses of multilayered dental ceramics subjected to biaxial flexure tests. *Dent. Mater.* 22: 460–469.

Hulm, B.J., Parker, J.D., and Evans, W.J. (1998). Biaxial strength of advanced materials. *J. Mater. Sci.* 33: 3255–3266.

ISO Standard 6872 (2015). *Dentistry – Ceramic Materials International Organization for Standardization*. Geneva: Switzerland.

Jadaan, O.M. and Wereszczak, A.A. (2009). *Revisiting the Recommended Geometry for the Diametrally Compressed Ceramic C-Ring Specimen*. ORNL-TM-2009/090. Oak Ridge, TN: National Laboratory (UT-Battelle, LLC & DOE) Cont: DE-AC05-00OR22725.

Jadaan, O.M., Shelleman, D.L., Conway, J.C. Jr. et al. (1991). Prediction of the strength of ceramic tubular components. I. analysis. *ASTM J. Test. Eval.* 19 (3): 181–191.

Jadaan, O.M., Shelleman, D.L., and Tressler, R.E. (1992). Lifetime prediction of internally pressurized siliconized SiC tubes subjected to creep rupture loading. *J. Am. Ceram. Soc.* 75 (2): 424–431.

Jeong, S.M., Park, S.E., and Lee, H.L. (2002). Fracture behaviour of alumina ceramics by biaxial ball-on-3-ball test. *J. Eur. Ceram. Soc.* 22: 1129–1135.

Klemensø, T., Lund, E., and Sørensen, B.F. (2007). Optimal shape of thin tensile test specimen. *J. Am. Ceram. Soc.* 90 (6): 1827–1835.

Lau, A.T.C. (2017). Why 30, a consideration for standard deviation. *ASTM Stand. News* July/August: 47–48.

Lu, C., Danzer, R., and Fischer, F.D. (2002). Fracture statistics of brittle materials: Weibull or normal distribution. *Phys. Rev. E* 65: 067102.

Mallick, P.K. (1993). *Fiber-Reinforced Composites: Materials, Manufacturing and Design*, 2e. New York: Marcel Dekker, Inc.

Marshall, D.B. (1980). An improved biaxial flexure test for ceramics. *Am. Ceram. Soc. Bull.* 59: 551–553.

Matthewson, M.J. and Kurkjian, C.R. (1987). Static fatigue of optical fibers in bending. *J. Am. Ceram. Soc.* 70 (9): 662–668.

Matthewson, M.J., Kurkjian, C.R., and Gulati, S.T. (1986). Strength measurement of optical fibers in bending. *J. Am. Ceram. Soc.* 69 (11): 815–821.

Medrano, R.E. and Gillis, P.P. (1987). Weibull statistics: tensile and bending tests. *J. Am. Ceram. Soc.* 70 (101): C-230–C-232.

Morrell, R., McCormick, N.J., Bevan, J. et al. (1999). Biaxial disc flexure – modulus and strength testing. *Br. Ceram. Trans.* 98: 234–240.

Nelson, P.R., Coffin, M., and Copeland, K.A.F. (2003). *Introductory Statistics for Engineering Experimentation*. New York: Elsevier.

Pashley, D.H., Carvalho, R.M., Sano, H. et al. (1999). The microtensile bond test: a review. *J. Adhes. Dent.* 1 (4): 299–309.

Pohanka, R.C., Freiman, S.W., and Bender, B.A. (1978). Effect of the phase transformation on the fracture behavior of $BaTiO_3$. *J. Am. Ceram. Soc.* 61 (1-2): 72–75.

Quinn, G.D. (2003). Weibull effective volumes and surfaces for cylindrical rods loaded in flexure. *J. Am. Ceram. Soc.* 86 (3): 475–479.

Quinn, G.D. (2016). *Fractography of Ceramics and Glasses*, NIST Special Publication 960-16e2. Gaithersburg, MD: NIST.

Quinn, G.D. and Morrell, R. (1991). Design data for engineering ceramics: a review of the flexure test. *J. Am. Ceram. Soc.* 74 (191): 2037–2066.

Quinn, G.D., Fuller, E.R., Xiang, D. et al. (2005). A novel test method for measuring mechanical properties at the small-scale: the theta specimen. In: *Mechanical Properties and Performance of Engineering Ceramics and Composites: Ceramic Engineering and Science Proceedings*, Ceramic Engineering and Science Proceedings, vol. 26, 117–126. Westerville, OH: American Ceramic Society.

Ritter, J.E. Jr., Jakus, K., Batakis, A., and Bandyopadhyay, N. (1980). Appraisal of biaxial strength testing. *J. Noncyst. Solids* 38-39: 419–424.

Robinson, P., Wilson, C., and Mecholsky, J. Jr. (2014). Processing and mechanical properties of hydroxyapatite-polysulfone laminated composites. *J. Eur. Cer. Soc.* 34 (5): 1387–1396.

Sadek, F.T., Muench, A., Poiate, I.A. et al. (2010). Influence of specimens' design and manufacturing process on microtensile bond strength to enamel – laboratory and FEA comparison. *Mater. Res.* 13 (2): 253–260.

Salem, J.A. and Powers, L.M. (2003) Guidelines for the testing of plates. In: *Proceedings of the 27th International Cocoa Beach Conference on Advanced Ceramics and Composites: B* (ed. W.M. Kriven and H.T. Lin). Ceramic Engineering and Science Proceedings, 24 (4): 357–364. Westerville, OH: The American Ceramic Society.

Semjonov, S. and Glaesemann, G.S. (2007). Chapter 18. High-speed tensile testing of optical fibers – new understanding for reliability prediction. In: *Micro- and Opto-Electronic Materials and Structures: Physics, Mechanics, Design, Reliability, Packaging*, vol. 1 (ed. E. Suhir, Y.C. Lee and C.P. Wong), A595–A625. New York: Springer.

Seshadri, S.G. and Chia, K.Y. (1987). Tensile testing of ceramics. *J. Am. Ceram. Soc.* 70 (10): C-242–C-244.

Shelleman, D.L., Jadaan, O.M., Conway, J.C. Jr. et al. (1991). Prediction of the strength of ceramic tubular components. II. Experimental verification. *ASTM J. Test. Eval.* 19 (3): 192–200.

Shetty, D.K., Rosenfield, A.R., Duckworth, W.H., and Held, P.R. (1983). A biaxial-flexure tests for evaluating ceramic strengths. *J. Am. Ceram. Soc.* 66: 36–42.

Soares, C.J., Soares, P.V., Santos-Filho, P.C., and Armstrong, S.R. (2008). Microtensile specimen attachment and shape – finite element analysis. *J. Dent. Res.* 87 (1): 89–93.

Spiegel, M.R., Schiller, J., and Srinivasan, R.A. (2009). *Schaum's Outline of Probability and Statistics*, 3e. New York: McGraw-Hill.

Timoshenko, S. and Woinowsky-Krieger, S. (1959). *Theory of Plates and Shells.* New York: McGraw-Hill Inc.

Vitman, F.F. and Pukh, V.P. (1963). A method for determining the strength of sheet glass. *Ind. Lab.* 29 (7): 925–930.

Wachtman, J.B. (1996). *Mechanical Properties of Ceramics.* New York: Wiley.

Wachtman, J.B., Capps, W., and Mandel, J. (1972). Biaxial flexure tests of ceramic substrates. *J. Mater.* 7 (2): 188–194.

Weibull, W. (1939). A statistical theory of the strength of materials. *Ingeniors Vetensk. Akad. Handl.* 151–153: 45–55.

Young, W.C. and Budynas, R. (2001). *Roark's Formulas for Stress and Strain*, 7e. New York: McGraw-Hill Professional.

CHAPTER 6

Thermally Induced Fracture

BACKGROUND

As noted in Chapter 1, not only mechanical loads but also temperature differentials can produce stresses that can lead to degradation in strengths or worse to sudden fracture. According to Kingery (1955), the first publication describing thermal stress-induced failure, commonly known as thermal shock, in ceramics is that of Winkelmann and Schott (1894). Since then there have been numerous attempts to develop both the physical and mathematical relations that could be used to predict thermal stress failure, as well as the test procedures needed to determine such susceptibility (Buessem 1955; Kingery 1955; Hasselman 1969; Becher 1981; Lewis 1983).[1] As we will show, expressions relating fracture susceptibility to temperature differences, cooling rates, and material properties have been derived, but because of uncertainties in the values of the properties particularly with respect to their temperature dependence, implementation of such predictive procedures is quite difficult.

[1] The entire issue 1, 1938 of the *J. Am. Ceram. Soc.* is devoted to thermal shock testing.

The Fracture of Brittle Materials: Testing and Analysis, Second Edition.
Stephen W. Freiman and John J. Mecholsky, Jr.

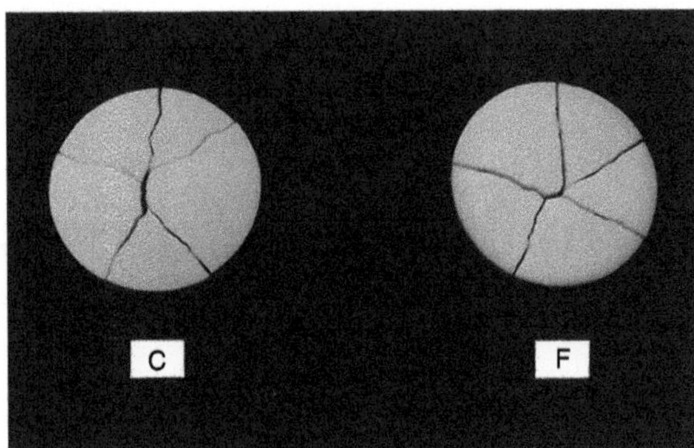

Figure 6.1 Typical crack pattern produced by thermal stress failure. Source: Tomba-Martinez and Cavalieri (2002). Reproduced with permission of John Wiley & Sons.

Thermally induced fractures can be identified by their fracture pattern. Thermal stress fractures in a part produce wavy cracks as shown in Figure 6.1. Thermal heating is frequently not uniform; hence, the stresses created by the heating and cooling are variable throughout the object. When a crack begins to propagate, it grows into a nonuniform stress field and tends to follow the maximum tensile stress regions that are changing with time. The local stress fields and the far-field stresses compete to guide the crack so that the crack generally follows a wavy or curved line.

GENERATION OF THERMAL STRESSES

It is a rapid change in temperature that is the major source of stresses that can lead to both instantaneous failure and a decline in strength, the phenomenon known as "thermal shock." The expansion or contraction of a portion of a solid body relative to another is the cause of the stresses. Consider a ceramic plate cooled rapidly from an elevated temperature. The atomic bonds at the surface of the plate are attempting to contract during cooling, but are restrained by the relatively hotter material in the interior. The stress at any point in the plate is determined by the difference in temperature between that point and the average temperature of the plate. The stresses in the plate are given by Eq. (6.1) (Kingery 1955):

$$\sigma_y = \sigma_z = \frac{E\alpha}{1-\mu}(T_e - T) \tag{6.1}$$

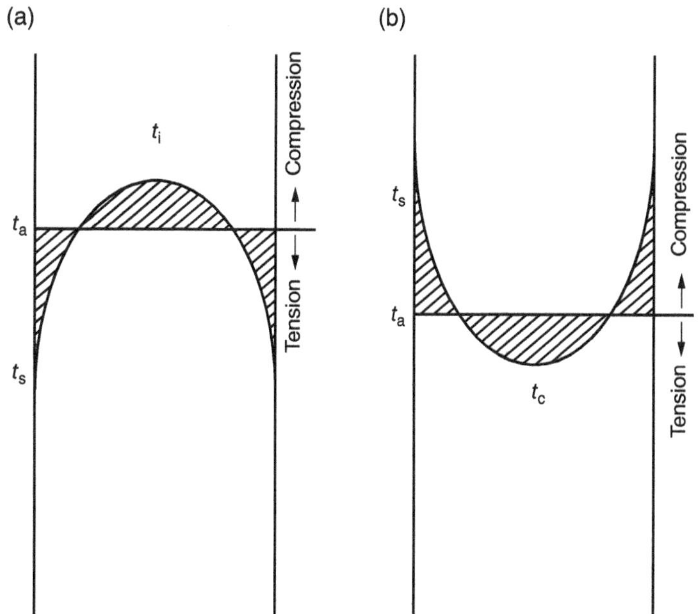

Figure 6.2 Temperature and stresses in a plate after (a) cooling and (b) heating. Source: Kingery (1955). Reproduced with permission of John Wiley & Sons.

where σ_y and σ_z are the stresses parallel to the dimensions of the plate at some depth into the thickness, E is Young's modulus, α is the thermal expansion coefficient, μ is Poisson's ratio, T_e is the temperature from which the plate was cooled, and T is the temperature at some point in the plate thickness at a given time. The distribution of stresses upon both cooling and heating a plate is shown in Figure 6.2.

Table 6.1 lists similar expressions for other shapes (Kingery 1955). A thermal shock resistance parameter, R, is obtained by rearranging Eq. (6.1) to give

$$R = \Delta T_C = \frac{\sigma_f(1-\mu)}{E\alpha} \tag{6.2}$$

Because fracture in most ceramics originates from the surface, cooling rapidly from elevated temperatures, thereby producing surface tensile stresses, is usually the most deleterious condition. This is why glass and ceramic parts are cooled slowly, so as to allow time

Table 6.1
Surface and Center Stresses for Various Shapes

Shape	Surface	Center
Infinite slab	$\sigma_x = 0$ $\alpha_v = \sigma_x = \frac{E\alpha}{(1-\mu)}(t_a - t_a)$	$\sigma_s = 0$ $\sigma_v = \sigma_x = \frac{E\alpha}{(1-\mu)}(t_\sigma - t_c)$
Thin plate	$\sigma_v = \sigma_x = 0$ $\sigma_x = \alpha E(t\alpha - t_\sigma)$	$\sigma_N = \sigma_t = 0$ $\sigma_s = \alpha E(t_\alpha - t_\sigma)$
Thin disk	$\sigma_r = 0$ $\sigma_\theta = \frac{(1-\mu)E\alpha}{(1-2\mu)}(t_\alpha - t_c)$	$\sigma_s = \alpha E(t_\alpha - t_\sigma)$ $\sigma_r = \frac{(1-\mu)E\alpha}{2(1-2\mu)}(t_\alpha - t_c)$ $\sigma_\theta = \frac{(1-\mu)E\alpha}{2(1-2\mu)}(t_\alpha - t_c)$
Long solid cylinder	$\sigma_r = 0$ $\sigma_\theta = \sigma_x = \frac{E\alpha}{(1-\mu)}(t_\alpha - t_\tau)$	$\sigma_\tau = \frac{E\alpha}{2(1-\mu)}(t_\alpha - t_\tau)$ $\sigma_\theta = \sigma_x = \frac{E\alpha}{2(1-\mu)}(t_\alpha - t_\tau)$
Long hollow cylinder	$\sigma_r = 0$ $\sigma_\theta = \sigma_x = \frac{E\alpha}{(1-\mu)}(t_\alpha - t_\tau)$	$\sigma_r = 0$ $\sigma_\theta = \sigma_x = \frac{E\alpha}{(1-\mu)}(t_\alpha - t_\tau)$
Solid sphere	$\sigma_r = 0$ $\sigma_t = \frac{E\alpha}{(1-\mu)}(t_\alpha - t_\tau)$	$\sigma_t = \sigma_\tau = \frac{2E\alpha}{3(1-\mu)}(t_\alpha - t_\tau)$
Hollow sphere	$\sigma_r = 0$ $\sigma_t = \frac{\alpha E}{(1-\mu)}(t_\alpha - t_\tau)$	$\sigma_r = 0$ $\sigma_t = \frac{\alpha E}{(1-\mu)}(t_\alpha - t_\tau)$

Source: Kingery (1955). Reproduced with permission of John Wiley & Sons.

for equilibration between the center and the surface. One must also take into account the heat transfer conditions at the surface of the material. This is accomplished by modifying Eq. (6.1) through the use of the Biot number, β, named after the French physicist Jean-Baptiste Biot:

$$\beta = \frac{a \bullet h}{k} \tag{6.3}$$

where a is a characteristic heat transfer length, typically the smallest dimension of the body; h is the surface heat transfer coefficient; and k is the thermal conductivity. Both thermal conductivity and the heat transfer coefficient can vary with temperature. In addition, h will vary with the medium into which the body is quenched and whether the surface of the material is rough or smooth. Other variables can include the incidence of nucleate boiling at the surface, heat lost during transfer out of the furnace, and even the rate at which the part falls through the quench medium.

Equation (6.1) now becomes

$$\sigma_y = \sigma_z = \frac{E\alpha}{1-\mu}\left(T_e - T\right)\beta \tag{6.4}$$

There are two potential outcomes when a part (or a specimen) is subjected to thermal stresses. The part could fail immediately, or it could be weakened due to the growth of flaws in the surface. The part will fracture at a stress, σ_f, whose value is determined by the critical temperature range over which the part or specimen is quenched as well as the quenching environment. The expression governing fracture is given by

$$\sigma_f = \frac{E\alpha}{1-\mu}\Delta T_C \beta \tag{6.5}$$

where ΔT_C is then the critical temperature differential during quenching. Because of the statistical nature of brittle fracture, there will be a range of strengths observed. In the interests of reliability, one should choose a lower limit to the fracture stress and determine the uncertainty in this value based on the methodology given in Chapter 8.

One can calculate the critical temperature difference ΔT_C by rearranging Eq. (6.5):

$$\Delta T_C = \frac{\sigma_f\left(1-\mu\right)}{E\alpha\beta} \tag{6.6}$$

$$\Delta T_C = \frac{\sigma_f(1-\mu)k}{E\alpha ah} \tag{6.7}$$

The difficulty in applying Eq. (6.7) to the prediction of thermal shock resistance lies in our lack of knowledge of the material parameters, particularly thermal conductivity and the surface heat transfer coefficient, h, which is highly dependent on the properties of the material and temperature, as well as the medium into which the body is quenched and the surface condition of the part or specimen.

In principle, if one knew all of the parameters in Eq. (6.7), then the critical quench conditions could be predicted. It is unlikely however that all of the material parameters will be available, especially since some of them will contain an unknown temperature dependence.

Collin and Rowcliffe (2000) used a fracture mechanics approach to develop an expression for the prediction of thermal shock resistance. Since it is valid to assume that normal flaws formed in the surface of a ceramic by machining or finishing are also accompanied by a residual stress, their analysis can be applied to any machined surface. The authors point out that the total stress intensity on such a crack subjected to quenching is the sum of the thermal and residual stress intensities:

$$K_{\text{total}} = K_{\text{thermal}} + K_{\text{residual}} \tag{6.8}$$

As noted in an earlier chapter, stress intensities for a semielliptical crack are largest at the deepest point of the crack. The residual stress intensity due to the indentation is given by the following expression (Anstis et al. 1981):

$$K_{\text{residual}} = \chi_{\text{surface}} P c^{-1/2} \tag{6.9}$$

where χ is a constant associated with the geometry of the crack (Anstis et al. 1981) and is a coefficient related to the residual stress state and c is the crack depth. If we combine Eqs. (6.9) and (6.10) and use the relationship between K and σ_f, the fracture strength, Collin and Rowcliffe (2000) developed an expression for the critical quenching temperature difference:

$$\Delta T_C = \frac{\sigma_f(1-\upsilon)}{E\alpha}\left(1.5 + \frac{B}{\beta} - 0.5e^{C/\beta}\right) \tag{6.10}$$

B and C are shape factors dependent on the specimen or part geometry.

For the purposes of this discussion, we will assume that the critical stress intensity is independent of crack size, i.e. no R-curve behavior. Then, Eq. (6.11) can be used to estimate the ΔT_C at which cracks will begin to grow.

THERMAL SHOCK TESTING

Because of the difficulty in predicting thermal stress resistance, it becomes necessary to perform tests on the particular material to determine its susceptibility to thermal shock. As seen in the following discussion, this can be a complex procedure and must be carefully tailored to the particular conditions under which the ceramic part is expected to operate. One should be aware that these tests provide only a relative guidance to the use of the material in different conditions.

The general approach to measuring thermal shock resistance in ceramics has been to quench flexural specimens from some elevated temperature into an ambient temperature bath, usually water (ASTM C1525-04 2013). Assuming that immediate fracture does not occur, the strength of the set of quenched specimens is then measured in flexure. The data is plotted as retained strength as a function of ΔT as shown in Figure 6.3. One should recognize that this test provides only a relative measure of thermal shock resistance. The susceptibility of any given material to either down-quenching or up-quenching is dependent on a wide variety of material and environmental properties.

One observes that at some critical ΔT, a reduction in strength takes place. Because of the statistical nature of fracture in these materials, one must be sure to test a sufficient number of specimens. One should also use the lower strengths as a guide to the safe operation of any component. The distribution in strengths of quenched specimens increases at the temperature differential at which a sharp decrease in strength is observed. The use of larger specimens and employment of test that stress a larger area, e.g. biaxial loading, will lead to a reduction in the breadth of the statistical distribution. One should recognize that quench testing provides only a relative measure of fracture resistance; each temperature and environmental condition could lead to different results.

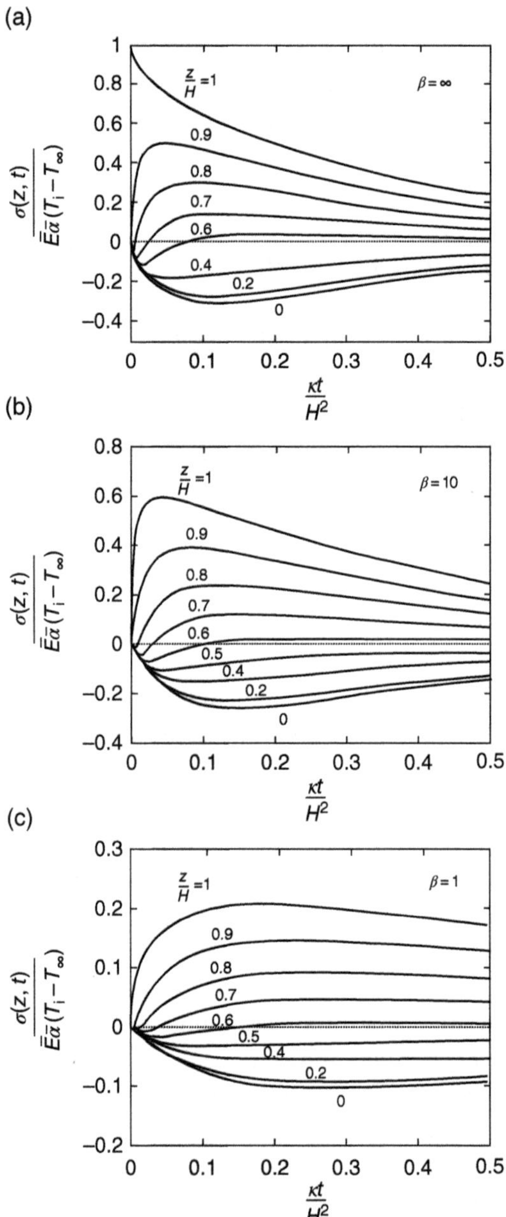

Figure 6.3 Evolution of dimensionless stress as a function of dimensionless time at selected locations for (a) $\beta=\infty$, (b) $\beta=10$, and (c) $\beta=1$ (Lu and Fleck 1998). Reproduced with permission of Elsevier.

There are two conditions that should be considered in conducting quench tests, the first in which the Biot number is quite large and the other in which it is low. When β is very small, there is almost complete thermal insulation, and thermally induced stresses are very low. The significant concern occurs when β is large, in which case the maximum stresses develop quickly. Lu and Fleck (1998) analyzed these situations as shown in Figure 6.4, which also includes an intermediate value of β. In Figure 6.3, dimensionless stress σ_d is plotted as a function of dimensionless time T_d:

$$\sigma_d \equiv \frac{\sigma}{E\alpha\left(T_i - T_\infty\right)} \tag{6.11}$$

where T_∞ is the temperature of the medium into which the material is quenched:

$$T_d \equiv \frac{kt}{H^2} \tag{6.12}$$

where H is the half thickness of the specimen.

As is evident from Figure 6.4, when quenched into a cold medium, the surface of the material is subjected to the largest tensile stress, which increases with increasing β. As a corollary, during hot shocking, the center of the specimen will experience the largest tensile stress.

Lu and Fleck (1998) also note that for small Biot numbers ($\beta < 1$), the most thermal shock-resistant materials will have a large value of $k\sigma_f/E\alpha$. However, when β is large, i.e. high surface heat transfer, better materials will have large values of $\sigma_f/E\alpha$. A plot of these conditions for a number of materials is shown in Figure 6.3.

They point out that the thermal diffusivity of the material is typically the controlling factor in determining the stress state.

Hasselman (1969) was the first to analyze thermal stress fracture from the perspective of both strength and fracture energy. As illustrated in Figure 6.4, he showed that there are regions of crack stability as well as regimes in which the strength is expected to decrease precipitously.

The latter had been demonstrated earlier by Davidge and Tappin (1967) as shown in Figure 6.5.

Typical experimental approaches to determining susceptibility to thermal shock involve heating a set of specimens in a furnace to a

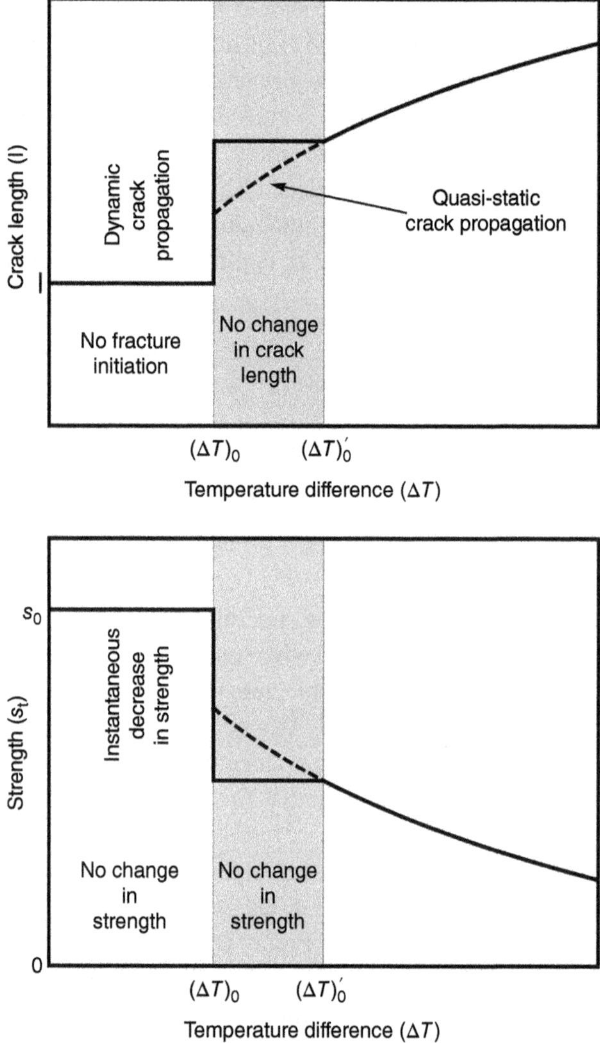

Figure 6.4 Crack length and strength as a function of thermal history. Source: Hasselman (1969). Reproduced with permission of John Wiley & Sons.

specific temperature and then quenching these specimens by dropping them into a liquid, typically water at room temperature. It has been argued that it is better to quench into water at 35 °C because of consistency of results (Lewis 1983). If quenching from very high temperatures,

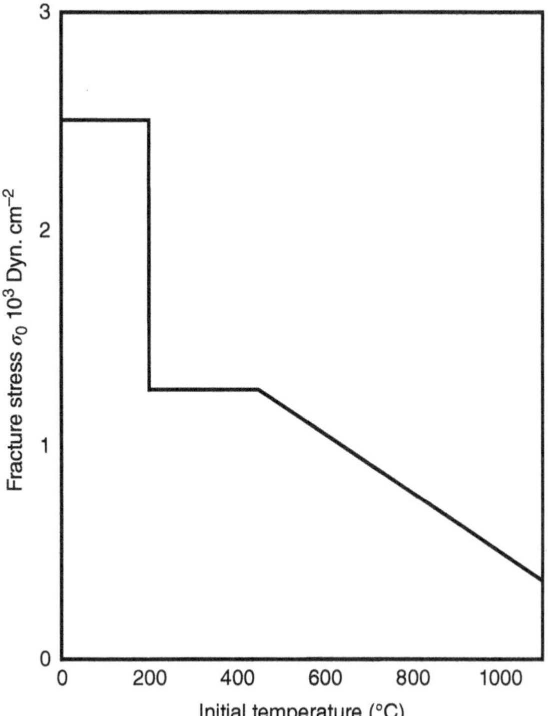

Figure 6.5 Room temperature strength of alumina rods subjected to thermal shock by quenching into water. Source: Davidge and Tappin (1967). Reproduced with permission of John Wiley & Sons.

e.g. >800 °C, one must be cautious to avoid radiant cooling of the specimens as they are removed from the furnace. Most data reported in the literature have been obtained using uniaxial flexure specimens tested in either three- or four-point bending (Lewis 1983).

Collin and Rowcliffe (2000) performed quenching experiments on four different materials that had been indented with a Vickers indenter following the procedure published by Andersson and Rowcliffe (1996). This indentation procedure is particularly useful in that it eliminates the need to perform fairly extensive flexural tests. Andersson and Rowcliffe (1996) placed indentations at several locations in polished plates of the materials, measured initial crack lengths, and quenched the plates from a range of temperatures. They defined the critical ΔT as that at which the average extension of the cracks was more than 10% of the original

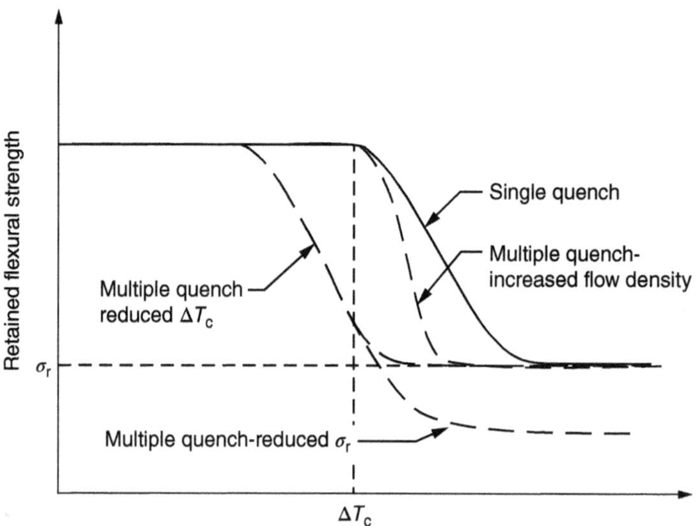

Figure 6.6 Schematic of single- and multiple-cycle thermal shock on retained strength. Source: Lewis and Rice (1981). Reproduced with permission of John Wiley & Sons.

length, and more than 25% of the cracks propagated. The authors concluded that ΔT_C was independent of the position of the indentation on the plate. They also determined that an indentation load of 70 N was optimum for quenching studies. They also suggested that when investigating actual components, the indentation load should be adjusted to create cracks as close in size to the actual cracks in the surface of the finished part. Effects of repeated thermal cycling on crack extension can also be conducted in this way.

Lewis and Rice (1981) studied thermal shock fatigue in both monolithic ceramics and ceramic–ceramic particulate composites. The results of their investigation are shown schematically in Figure 6.6.

OTHER PROPOSED THERMAL SHOCK TESTS

Faber et al. (1981) proposed a variation in the quench experiment by devising an apparatus in which cold air under high pressure is impinged on a heated, previously indented, specimen (Figure 6.7). They note that this test has limitations in terms of the thickness of the specimen that can be used and the maximum temperature to which it can be raised.

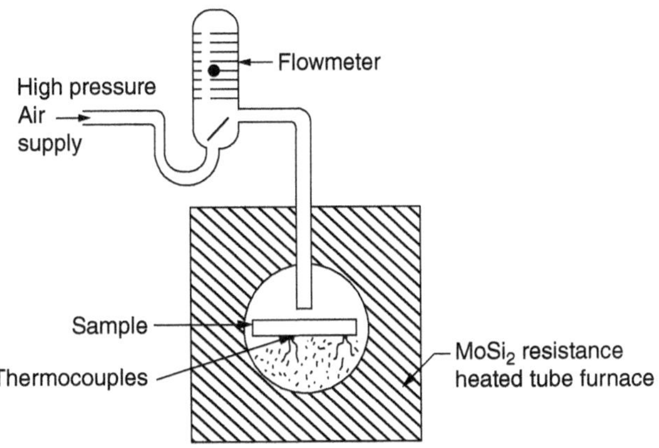

Figure 6.7 Schematic of thermal stress test apparatus (Faber et al. 1981).

As noted at the beginning of this chapter, thermal stresses are also generated during rapid heating of a material, possibly leading to failure from the interior, or if heating occurs only on one surface, from the rear surface of the part or specimen. Tests have been devised to investigate this possibility.

For example, Schneider and Petzow (1991) devised a test in which circular disks are heated up to 1350 °C using two tungsten halogen lamps. They used this test to measure fracture in silicon nitride. They claimed that the test could be used on "dark" materials such as Si_3N_4, SiC, and AlN as well as "white" ceramics such as Al_2O_3 and ZrO_2, provided that they are coated with a radiation-absorbing film.

Mecholsky et al. (1980) showed that continuous wave laser irradiation could be used to rank the thermal shock resistance of ceramics. They used time to failure as the ranking criterion. They showed that laser irradiation tests provided rankings similar to water quench, provided that geometry, degree of opaqueness, and heating rates are equivalent.

SUMMARY

In this chapter we have outlined both the physics and the testing procedures that can be used to determine the susceptibility of ceramics to thermal stress-induced failure. This is a field that is continually evolving, as new test methods and analysis procedures are being developed.

QUESTIONS

1. Describe how Biot's modulus is affected by a change in geometry for a thermally conducting material versus a thermally insulating material.
2. Sketch the crack growth, Δc, of a ceramic material as a function of temperature difference, i.e. ΔT, from ambient room temperature to the designated temperature as the temperature is increased. (The resulting graph will be Δc vs. ΔT.)
3. Why would the results of residual strength for a ceramic at the critical temperature difference, ΔT_C, be different if you quench in water versus oil?
4. Calculate the expected critical thermal shock temperature for a silicon nitride disk quenched in water. The properties are the following: thermal conductivity of 30 W $(m \cdot K)^{-1}$, coefficient of thermal expansion of $3.3 \cdot 10^{-6}$ per °C, fracture toughness of 6 MPa $\cdot$ $m^{1/2}$, Poisson's ratio of 0.27, and elastic modulus of 310 GPa. The disk is 10 mm × 10 mm × 1 mm. Assume that the disk has been polished and protected, so it has cracks <100 μm.
5. If you are requested to test the thermal shock resistance of siliconized silicon carbide hollow tubular components that are 1 m in length, 50 mm in outside diameter, and 10 mm thick, what test procedure would you recommend, and why? The tubes are used as heat exchangers in a high temperature furnace. How would you plan to ensure that your tests are representative of in-service behavior?
6. Materials that are susceptible to environmentally aided slow crack growth due to stress corrosion cracking (SCC) processes can also be susceptible to thermal shock. How will the SCC susceptibility affect thermal shock measurements?

REFERENCES

Andersson, T. and Rowcliffe, D.J. (1996). Indentation thermal shock test for ceramics. *J. Am. Ceram. Soc.* 79: 1509–1514.

Anstis, G.R., Chantikul, P., Lawn, B.R., and Marshall, D.B. (1981). A critical evaluation of indentation techniques for measuring fracture toughness: I, direct crack measurements. *J. Am. Ceram. Soc.* 64: 533–538.

ASTM C1525-04 (2013). *Standard Test Method for Determination of Thermal Shock Resistance of Advanced Ceramics*. West Conshohocken, PA: ASTM International.

Becher, P.F. (1981). Effect of water bath temperature on the thermal shock of Al_2O_3. *Comm. Am. Ceram. Soc.* 64: C17–C18.

Buessem, W.R. (1955). Thermal shock testing. *J. Am. Ceram. Soc.* 38: 15–25.

Collin, M. and Rowcliffe, D. (2000). Analysis and prediction of thermal shock in brittle materials. *Acta Mater.* 48: 1655–1665.

Davidge, R.W. and Tappin, G. (1967). Thermal shock and fracture in ceramics. *Trans. Brit. Ceram. Soc.* 66: 405–422.

Faber, K.T., Huang, M.D., and Evans, A.G. (1981). Quantitative studies of thermal shock in ceramics based on a novel test technique. *J. Am. Ceram. Soc.* 74: 296–301.

Hasselman, D.P.H. (1969). Unified theory of thermal shock fracture initiation and crack propagation in brittle ceramics. *J. Am. Ceram. Soc.* 52: 600–604.

Kingery, W.D. (1955). Factors affecting thermal stress resistance of ceramic materials. *J. Am. Ceram. Soc.* 38: 3–15.

Lewis, D. III (1983). Thermal shock and thermal fatigue testing of ceramics with the water quench test. In: *Fracture Mechanics of Ceramics*, 5e (ed. R.C. Bradt, A.G. Evans, D.P.H. Hasselman and F.F. Lange), 487–496. New York: Plenum Press.

Lewis, D. III and Rice, R.W. (1981). Thermal shock fatigue of monolithic ceramics and ceramic-ceramic particulate composites. *Ceram. Eng. Sci. Proc.* 2: 712–718.

Lu, T.J. and Fleck, N.A. (1998). The thermal shock resistance of solids. *Acta Mater.* 46 (13): 4755–4768.

Mecholsky, J.J., Becher, P.F., Rice, R.W. et al. (1980). Laser induced thermal stresses in brittle materials. In: *Thermal Stresses in Severe Environments* (ed. D.P.H. Hasselman and R.A. Heller), 567–589. New York: Plenum Publishing.

Schneider, G.A. and Petzow, G. (1991). Thermal shock testing of ceramics – a new testing method. *J. Am. Ceram. Soc.* 74: 98–102.

Tomba-Martinez, A.G. and Cavalieri, A.L. (2002). Fracture analyses of alumina subjected to mechanical and thermal shock biaxial stresses. *J. Am. Ceram. Soc.* 85: 921–926.

Winkelmann, A. and Schott, O. (1894). Ueber thermische Widerstands-coefficienten verschiedener Glaser in ihrer Abhangigkeit von der chemischen Zusammensetzung. *Ann. Phys. Chem.* 51: 730.

CHAPTER 7

Modeling of Brittle Fracture

INTRODUCTION

Previously we presented the measurement techniques available to determine the fracture resistance of brittle materials, i.e. fracture energy or fracture toughness. It is clear, however, that in most instances fairly substantial quantities of material are required to conduct accurate tests. As noted in Chapter 3, the test used most frequently to determine the fracture toughness of small specimens, indentation crack length, has significant limitations. While the indentation-crack-length procedure has been used to determine fracture toughness of small crystals, the values of which compare favorably with those obtained in other ways (Raynes et al. 1991), the accuracy of such tests is questionable. It would be valuable if one could estimate the crack growth resistance of new materials of potential interest or already under development. In this chapter, we will attempt to describe some of the attempts that have been, and continue to be made, to predict fracture resistance based upon other material parameters, as well as from first-principle atomistic approaches.

The Fracture of Brittle Materials: Testing and Analysis, Second Edition.
Stephen W. Freiman and John J. Mecholsky, Jr.

MODELS BASED ON MATERIAL PROPERTIES

Gilman (1960) was one of the first to suggest that brittle fracture could be modeled using a simplistic approach. Basically, he equated the work necessary to separate cleavage planes in a crystal to the energy of the two new surfaces formed during fracture. He used a sine function to model the bonding between the fracture (cleavage) planes in the material. He estimated the strain between the surfaces as y/d_0, where y is the separation distance and d_0 is the equilibrium distance between the cleavage planes. By assuming the sine function equal to its argument at small strains, in order to determine Young's modulus, E, Gilman derived the following expression for surface, i.e. fracture, energy:

$$\Upsilon = \left(\frac{E}{d_0}\right)\left(\frac{a}{\pi}\right)^2 \tag{7.1}$$

where a is the separation distance at which bond failure is assumed to occur. For reasons not given in the publication, Gilman took the value of a to be "equal to the atomic radii of atoms lying in the cleavage planes" and noted that the calculated values of Υ compare favorably with measurements. However, we can determine no physical basis for his assumption.

Nonetheless, other investigators (Becher and Freiman 1978) also found a good correlation between this model and experimental results, but without a firm basis for the failure criterion, this approach lacks credibility.

More recently, Freiman and Mecholsky (2010) generalized Gilman's approach. They modeled the fracture process as two planes that are being separated by a tensile stress perpendicular to them. They assumed that the process is carried out such that no energy is lost to other processes, e.g. dislocation generation, heat, etc. Then the energy under a stress-separation curve is the fracture energy, Υ. The presence of a crack simply serves to concentrate stress.

The separation, δ, of two planes of atoms from their equilibrium value, d_0, is given by

$$2\gamma = U(\infty) - U(d_0) = \int_{d_0}^{\infty} \sigma(R)\,\mathrm{d}R \tag{7.2}$$

where $U(d_0)$ is the equilibrium energy of the crystal and $U(\infty)$ is the energy at complete separation. $R \equiv d_0 + \delta$, and $\sigma(R)$ is the particular restoring stress governing separation of the planes.

We define $\varepsilon \equiv \delta/d_0$. Then, under plane strain conditions,

$$\left.\frac{d\sigma}{d\varepsilon}\right|_{\varepsilon=0} = E \tag{7.3}$$

where E is Young's modulus, whose value is calculated perpendicular to the plane of fracture. For any $\sigma(R)$ we might choose, there exists a strain ε_m for which the restoring stress $\sigma_m = \sigma(\varepsilon_m)$ is a maximum.

As an example, we can use the sine function for a bonding expression. It has a well-defined integration limit, the point of failure. The sine law can be expressed as

$$\sigma(\varepsilon) = \sigma_m \sin\left(\frac{\pi\varepsilon}{2\varepsilon_m}\right) \quad \text{for } 0 \le \varepsilon \le 2\varepsilon_m \tag{7.4}$$

$$\sigma(\varepsilon) = 0 \qquad \text{for } 2\varepsilon_m \le \varepsilon \tag{7.5}$$

Then it can be shown that

$$E = \frac{\pi\sigma_m}{2\varepsilon_m} \tag{7.6}$$

Equation (7.2) then gives

$$2\gamma = \sigma_m d_o \int_0^{2\varepsilon_m} \sin\left(\frac{\pi\varepsilon'}{2\varepsilon_m}\right) d\varepsilon' = \frac{4\varepsilon_m \sigma_m d_o}{\pi} \tag{7.7}$$

Combining Eqs. (7.6) and (7.7) gives

$$\gamma = \frac{4}{\pi^2}\varepsilon_m^2 d_0 E \tag{7.8}$$

or

$$\gamma = k d_0 E \tag{7.9}$$

where k is a numerical constant, whose value depends on the particular stress–strain function. Integrating other expressions, e.g. the Born function, will lead to an equation of the same form with a different value of k.

Any stress–strain function must have the same basic features:

- An initial slope defined as Young's modulus, E.
- A maximum stress, σ_m, at some strain, ε_m.
- A parameter that defines the range over which the function operates.
- An area under the stress–strain curve ~ $2\gamma/d_0$.

Each function contains an unknown numerical parameter, either ε_m or a failure criterion. However, without detailed atomistic calculations, it is not possible to assign values to these parameters (Talman and Shadwick 1976). As noted above, while Gilman (1960) attempted to rectify this situation by assigning a value to the failure criterion, i.e. the period of the sine function, as being one-half of the radii of the atoms in the fracture plane, we can find no physical basis for this choice. Despite our lack of knowledge regarding the value of k, we can use Eq. (7.9) to compare models with experimental data.

Freiman and Mecholsky (2010) compared Eq. (7.9) with fracture experiments by choosing a set of materials having a common (cubic) crystal structure for which experimental fracture energy data was available. No attempt was made to differentiate between good and poor cleavage or to discriminate between selections based on the perceived quality of the test procedure or the experimental data. Values of the elastic constant pertinent to the fracture system for each material were calculated by rotating the $E = 1/S_{11}$ values reported in the literature (Simmons and Wang 1971) to the direction perpendicular to the fracture planes (Table 7.1). The equilibrium plane spacing, d_0, was calculated from the reported lattice parameter and the crystal structure for the relevant set of fracture planes (Table 7.1). When data was taken from the literature, reported values of fracture toughness (as opposed to fracture energy) were converted to γ through suitable expressions (Sih et al. 1965). It was assumed that the plane separation process was independent of any other material properties or structure.

The line in Figure 7.1 is a best linear fit to the data. Because, in the ultralow elastic constant regime, it was assumed that other types of bonding would apply, we did not force the fitted lines through the origin. Note that the type of atomic bonding in the materials ranging from completely ionic (e.g. KCl, CaF_2) to completely covalent (e.g. Si, Ge,

Table 7.1

Data Used in Comparison of Measured and Calculated Values of Fracture Energy

Material	Plane	d_0 (nm)	E (GPa)[a]	Υ (J m^{-2})	Reference
KCl	100	0.31	22.9	0.25	Freiman et al. (1975)
KCl	100	0.31	22.9	0.11	Westwood and Hitch (1963)
BaF	111	0.27	66.2	0.35	Becher and Freiman (1978)
ZnSe	110	0.25	78.6	0.60	Freiman et al. (1975)
CaF	111	0.24	88.4	0.51	Becher and Freiman (1978)
ZnTe	110	0.30	63.9	0.56	White et al. (1988)
GaAs	110	0.24	121.5	0.87	White et al. (1988)
GaP	110	0.27	144.6	0.94	White et al. (1988)
Ge	111	0.25	137.7	1.1	Jaccodine (1963)
Si	111	0.24	174.8	1.2	Chen and Leipold (1980)
MgO	100	0.21	335.9	2.4	Shockey and Groves (1968)
MgO	100	0.21	335.9	1.2	Westwood and Goldheim (1963)
MgO	100	0.21	335.9	1.2	Gutshall and Gross (1965)
$MgAl_2O_3$	100	0.19	363.8	3.3	Stewart and Bradt (1980)
Diamond	111	0.2	1207.0	10.0	Novikov and Dub (1991)

[a] Values of the elastic constant in each material were calculated by rotating the $1/S_{11}$ values reported in the literature Simmons and Wang (1971) to the direction perpendicular to the fracture planes.

diamond) does not appear to affect the fit of the data. There seems to be a single k common to the entire range of cubic materials. However, because k contains an unknown parameter (either the failure criterion or the point of maximum stress), this alone cannot be used to identify the stress–strain function that best applies. We also recognize that there are limitations to this analysis because of the lack of experimental data for cubic materials in the intermediate range of 4–10 J m^{-2}.

Note that because there are only slight variations in the lattice parameter among these materials, one can plot fracture energy versus Young's modulus with little loss in accuracy. It has been noted in other publications (Mecholsky et al. 1976) that there is a monotonic relationship between fracture energy or fracture toughness and elastic

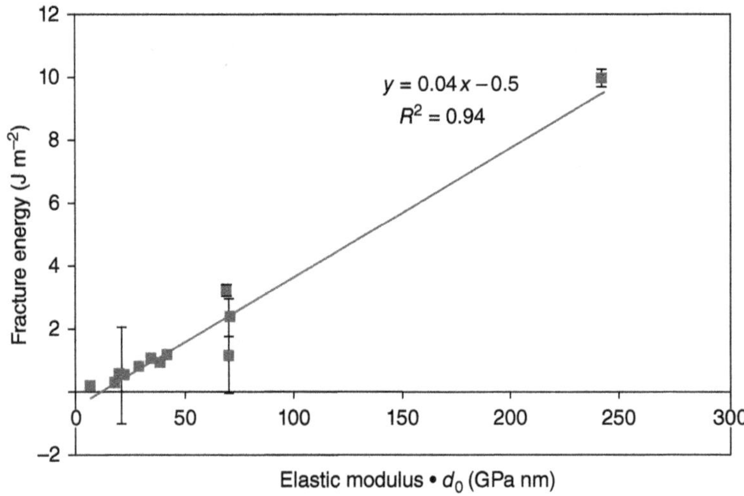

Figure 7.1 Measured fracture energies plotted versus $E \cdot d_0$ as suggested by Eq. (7.9).

modulus even with the complications introduced by the presence of microstructure. Yuan and Xi (2011) used a very similar approach to predict the strength of metallic glasses. They demonstrated a clear correlation between strength and elastic modulus in a wide variety of glass compositions. An interesting exception to this trend was reported by Kennedy et al. (1973) for a series of soda-silica glasses in which the percentage of Na_2O was varied from 0 to 35%. The authors reported that while the elastic modulus decreased with increasing soda content, the fracture toughness increased.

Note also that the form of a stress–strain (atomic bonding) function will not affect the relationship – as long as it applies equally to all of the materials – so that this approach cannot be used to distinguish between interatomic potentials. For example, Tromans and Meech (2002, 2004) used a Morse function to calculate the toughness of ionic solids and the Born model in the calculation for covalent materials. Also, nothing in the model allows for prediction of different fracture energies in different directions in a fracture plane.

King and Antonelli (2007) also attempted a simple bond energy approach to predicting fracture. They assumed that the fracture energy could be calculated as the energy needed to break all of the chemical bonds in a unit area of material. They suggest that this approach is applicable to thin film as well as bulk materials.

Tromans and Meech (2002, 2004) used several methods to calculate the fracture toughness of both covalent minerals and large variety of oxides, sulfides, silicates, and halides. Their interest was from a mineralogical point of view of predicting the fracturing of small particles.

Lazar et al. (2005) used the universal binding energy relation (UBER) (Rose et al. 1981) to model the fracture of brittle crystals. They assumed that the strain leading to fracture is localized in a small distance from the cleavage plane. They calculate a so-called localization length of 2.4A. The results of their calculations agree reasonably well with experimental values of fracture energy. Lazar and Podloucky (2008) used density functional theory (DFT) to calculate the energetics of cleavage. They modified the UBER model to account for relaxation in the atomic bonding during fracture.

Despite the relatively good correlation between some of these models and experimental fracture energy data, it is clear that both the fundamental assumptions underlying the calculations and the models themselves are too simplistic. An example of this issue is shown in Figure 7.2, in which fracture energy data for SiO_2, SiC, and Al_2O_3 crystals, having non-cubic crystals structures, are plotted along with the data shown previously for cubic materials as a function of elastic constant. It is obvious that the above materials do not behave in the same way as those with a cubic crystal structure.

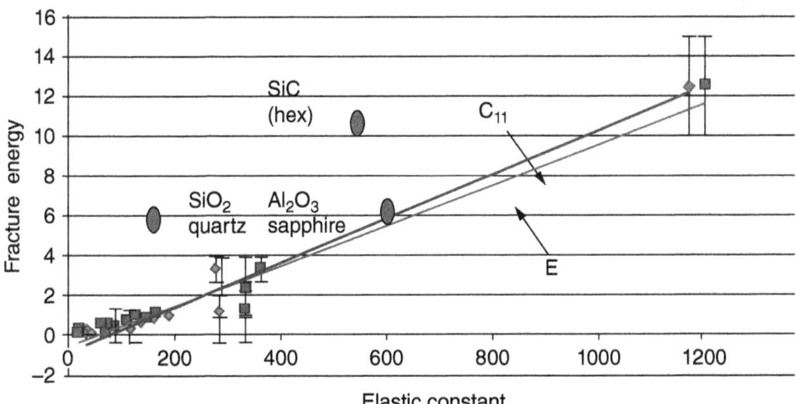

Figure 7.2 Calculated fracture energies of single crystals as a function of only elastic constant.

One could argue that fracture energy and elastic constant are still directly related but that some other factor changes the proportionality constant. One such parameter could be the number of atoms that are involved in the fracture process. As was noted in an earlier chapter, there is a region very near the tip of a crack under stress in which linear elasticity can no longer be assumed. The size of the zone and the atomic interactions within it could play a role in determining crack growth resistance. Such a zone may be the reason behind the trend in fracture toughness for the series of soda-silica glasses noted previously (Kennedy et al. 1973).

LATTICE TRAPPING AND ATOMISTICS OF FRACTURE

It was Thomson et al. (1971) who first suggested that the discreteness of a crystal lattice should lead to fracture behavior not expected from a continuum model. They termed this effect "lattice trapping." The presence of lattice trapping was expected to increase the energy required to move the crack over that predicted by continuum energy models. Nevertheless, at the time of the publication, they could not describe an observable experimental manifestation of this phenomenon.

Gumbsch and Cannon (2000) noted that the magnitude of the lattice trapping effect strongly depends on the force law governing the atomic interaction. They suggested that snapping spring forces will give rise to larger lattice trapping effects than "softer" type bonding. They note that a significant consequence of lattice trapping is anisotropy with respect to the direction of crack growth within a given cleavage plane.

Silicon has been the "go-to" material for many of the modeling studies of the atomistics of fracture. Silicon has two principal cleavage planes $\{111\}$ and $\{110\}$. Gumbsch and Cannon pointed out that directional anisotropy has been noted on both sets of planes. Margevicius and Gumbsch (1998) later observed anisotropy in cleavage energy of GaAs along different directions. A recent review article (Bitzek et al. 2015) gives an up-to-date perspective on the topic of the atomistics of fracture.

In addition to purely brittle fracture prediction, it would be quite valuable to be able to understand more fully the effect of external environments on crack propagation. As noted in Chapter 4, at present there is no a priori method of predicting susceptibility to environmentally enhanced fracture based on a material's chemical composition.

MODELING OF ENVIRONMENTAL INTERACTIONS

There have been attempts made to predict stress-dependent reactions of environments, e.g. water, with both silica and silicon. Following is a brief discussion of some of these approaches. Wong-Ng et al. (1992) used molecular orbital (MO) calculations to determine the effects of bond strain on the charge distribution in silica. They chose a Si—O fragment from a silica lattice as shown in Figure 7.3 and employed a program (Gaussian90) to determine effects of both bond angle variations and bond stretching – up to 20% strain – on both the total electron distribution and bonding electron distribution on both the silicon and oxygen atoms.

Wong-Ng et al. note that while the absolute value of the electron distribution depends on the exact configuration of the strain, the general trends remained the same. They point out that while the total number of electrons on the oxygen atom increases with strain, the number of bonding electrons decreases. Both the total number of electrons and the bonding electrons on the silicon atom decrease with strain. They conclude that this result is in agreement with the prediction of the model put

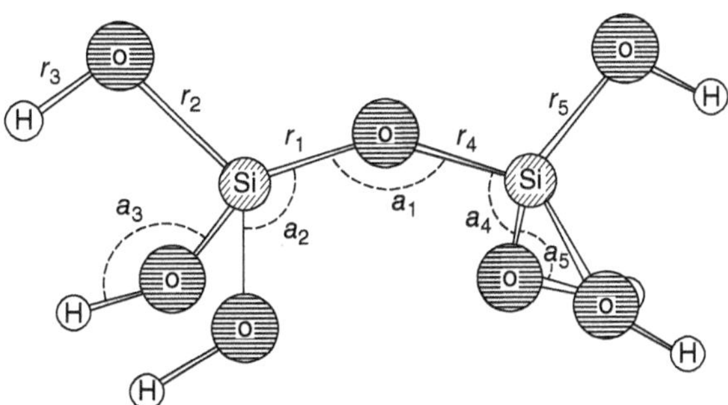

Figure 7.3 Schematic of silica fragment used in the MO calculations. The hydrogen atoms are inserted for charge compensation. Source: Wong-Ng et al. (1992). Reproduced with permission of John Wiley & Sons.

forward by Michalske and Freiman (1983), which suggests that strain at the crack tip makes the bonds more reactive. In a second part of the study (Lindsay et al. 1994), the same MO approach was used to examine the effect of applying stress to the Si—O bond in the presence of five environments – ammonia, water, formamide, nitrogen, and argon. The authors conclude that it is the energy required to move a molecule toward the strained bond that is the most important factor in the reaction process. Lindsay et al. also conclude that steric hindrance plays a role as well. It should be noted that these calculations do not take temperature into account.

Wong-Ng et al. (1996) performed similar MO calculations on strained silicon. They concluded that silicon showed no tendency to charge polarization due to strain and that straining a Si—Si bond does not lead to any attractive force between the bond and a water molecule. This result agrees with existing experimental data as discussed in Chapter 3. Note that the argument against the use of MO calculations is that it includes too few atoms that may participate in the fracture process.

Del Bene et al. (2003) carried out a quantum mechanics calculation on the reaction of water with silica using second-order perturbation theory [MBPT(2)] with the 6-31+ G(d,p) basis set. Their calculations suggest that it is a water molecule dimer rather than a monomer that would react with the strained Si—O—Si bond. Another paper along the same lines (Zhu et al. 2005) discusses potential molecular pathways for the silica–water reaction.

Amorphous materials introduce a somewhat different issue. West and Hench (1994a) used a semiempirical method based on Hartree–Fock bonding to model the fracture of silica rings (Bell and Dean 1972). While a drawback of semiempirical techniques is their reliance on experimental data, they can be used to model much larger groups of atoms. Part I of West and Hench series of papers describes the reaction pathway for fracture of polysiloxane rings in the absence of any external environment. West and Hench calculate the transition state barriers to ring fracture for a series of ring geometries and suggest that a crack tip in silica will move along the lowest energy pathways. They calculate that fracture by ring contraction in the absence of water has an activation barrier of 77 kcal mol^{-1} for fourfold rings compared with

96 kcal mol^{-1} for threefold and 103 kcal mol^{-1} for fivefold rings. Based upon this activation barrier data, West and Hench suggested that fracture of fourfold rings should be expected in region III of the crack growth curve.

The second paper in this series (West and Hench 1994b) examines the effects of water in influencing the path of a crack through a silica ring structure. Again using activation energies as a guide, West and Hench concluded that in the presence of water, threefold rings will be the primary site at which bond rupture will occur; i.e. cracks will seek out threefold ring structures to follow as they grow. In the third paper in the series (West and Hench 1995), the authors discuss fracture of five- and six-member rings.

It is not necessary that the entire crack front move simultaneously through the structure. When portions of the crack front encounter weak rings, so-called kinks can form. The energy of these "kinks" is such that they can move parallel to the front and so cause the crack to move forward. The concept of "kinks" as the sites of enhanced incidence of bond rupture was put forward by Lawn (1975) in an explanation of the fracture process.

Molecular dynamics (MD) is computer approach to modeling the fracture process, in which one can follow the simulated motion of the atoms or molecules of material. MD solves Newton's equations of motion for a set of particles. MD has been used fairly extensively to explore the structure and the brittle fracture process in glasses (Muralidharan et al. 2005). Both static and dynamic properties of the system can be evaluated as a function of temperature. A primary requirement is an accurate representation of the interatomic potential between the entities. It is outside the scope of this chapter to discuss the numerous potentials that have been tried to simulate the silica structure. The publication by Muralidharan et al. (2005) provides an excellent review of the field of MD simulation of silica fracture up until that time. There continue to be attempts to use MD to describe fracture. For example, Tang (2009) reported on an MD study of the effect of water on the fracture of cristobalite and claimed good agreement between the model and experimental values of fracture toughness. A significant limitation is the inability to, a priori, establish an atomic bonding potential for the material of interest.

A NEW APPROACH TO MODELING: APPLICATION OF FRACTAL GEOMETRY

The simulation of fracture should be able to help answer several questions. Once a crack starts to grow, in what manner does it propagate? Is the fracture surface smooth? Is it tortuous? Why is a fracture surface rough at several length scales? Is it possible to predict fracture toughness from a material's structure? At the present time, fracture experiments cannot answer these questions directly due to the speed and scale at which bond breaking occurs. As noted earlier, calculations of fracture using various methods are also limited. Without physical experiments to compare with results of modeling, the determination of material properties is only suggestive, not predictive. In other words, due to the nature of atomic simulations, the extrapolation to the macroscopic scale is very difficult without a methodology for scaling.

What is missing is a property that can be detected quantitatively in both simulations and physical experiments. One of the unaddressed attributes of fracture, the tortuosity of the fracture surface, is suspected to transverse length scales in a predictable manner. Roughness is a measure of the deviations from a two-dimensional (2-D) surface or some partial dimension greater than 2-D. Roughness is usually measured using one stylus. However, if several styli are used in measuring a surface, we can measure the tortuosity at many length scales by determining the fractal dimension. This partial dimension (or fractal dimension) is a construct that exhibits self-similarity and scale invariance. Self-similarity means that one region of the fracture surface will appear to have statistically the same shape as another region of the fracture surface. Scale invariance means that the fracture surface will appear the same at many magnifications, i.e. unless a scale marker is in the micrograph, the observer will not be able to identify the actual size of the object observed on the fracture surface. Sometimes the scaling factor may be anisotropic, that is, the magnification in one direction is different from another direction. These surfaces are called self-affine. Fracture surfaces are mostly self-similar but can be self-affine.

The fractal dimension has been demonstrated to be quantitatively related to the fracture toughness, providing a measurable connection between crack propagation and resistance to fracture (cf. Chapter 6), implying that the fracture path is dependent on the arrangement of atoms and strength of the bonds connecting them. Based on this premise, a simulation that produces a fracture surface with the same fractal dimension and same fracture toughness as an experimental analog can be taken as a demonstrative fracture event allowing for a detailed analysis of crack propagation. Such a simulation would expand the practical applications of fracture theory to materials that are not adequately handled by stress intensity factor consideration. For example, if the fractal dimension can be quantitatively related to bond strength or even the sequence in which bonds break, the fracture toughness of a material could be predicted purely from a simulation. The incentive is a framework to determine the fracture toughness of new materials without actually fabricating them.

To this end, there are several experimental and computational questions that need to be answered. All of these answers are related to the ability to determine the validity of a computational study of fracture relative to experiment. Fracture is a fundamentally multi-scale problem; a global stress induces a local stress causing atomic bonds to break, which transcends to multiple bond breaking, leading to a very dynamic fracture process that spans the atomic to the macroscopic. It may be eventually possible to capture all these processes in a single computational method, but this is not the case at the moment. A pragmatic approach would be to separate a fracture event into constitutive processes that are handled by tailored methods. This would require experiments designed for the limitations of computational simulations and corresponding quantitative experimental measurements. Fractography provides a set of such measurements; while not all inclusive, they are necessary. By measuring the fractal dimension of the fracture surface, a quantitative value can be assigned to the surface. This value is related to the fracture energy of the fracture process. Thus, any successful model should be able to predict the fractal dimension and the fracture energy. What is missing is a connection to the atomic or molecular length scale. In order to supply this missing knowledge, we consider the fracture process in terms of the fractal concept.

Fractal geometry is a mathematical construct that allows for non-integer dimensions. It is characterized by self-similar (or self-affine) structures that are scale invariant. Fracture surfaces are generally self-similar and in some cases self-affine (Bouchaud et al. 1990, 1993), scale invariant, and characterized by their fractal dimension, D (Mandelbrot et al. 1984). The fractal dimension is composed of the topological dimension, i.e. 1, 2, or 3, and the incremental dimension above the topological dimension, e.g. 2.3, represents a plane with a tortuous surface and would be indicated by $2.D^*$ ($D^* = 0.3$). Fundamental relationships can be derived between the fractal dimensional increment, D^*, and the fracture energy, γ:

$$2\gamma = Ea_0 D^* \tag{7.10}$$

where E is Young's modulus and a_0 is a parameter having the units of length (Mecholsky et al. 1989; Russ 1994; West et al. 1999; Hill et al. 2001). The discussion of the meaning of a_0 is a key element and will be discussed in detail later. However, it is the necessary link to the atomic and molecular length scale that has been missing. The relationship between γ and D^* is based on experimental observations (Figure 7.4). The measurement of D^* is an average property of the entire fracture surface and a measure of its tortuosity.

The fractal dimensional increment (D^*) and a_0 are two experimental parameters at different length scales: a surface (~millions of atoms) and a local cluster. These parameters can be transformed into criteria that determine the fracture toughness, identify the manner of crack propagation, and provide a means to analyze how local bonding influences macroscopic fracture. Determining either D^* or a_0 computationally is a significant challenge. Such calculations can be carried out, but their validity is currently hard to qualify. And this is because bond breaking, especially of large clusters, is still a difficult and sometimes unsolvable problem even with the most rigorous quantum calculations.

Previous MD models of fracture in amorphous silica and single crystal silicon have shown that fracture can be reasonably explained by the creation of free volume in the structure and is chaotic in nature. From these simulations, Swiler et al. (1995) suggested that the fracture surface results from the linking of regions of free volume in amorphous materials. The simulations were compared with experimental results (Tsai 1993; Tsai et al. 1997) by constructing simulated fracture surfaces from the computer models and generating horizontal contours. From the perimeter of the contours, a fractal dimension was determined using

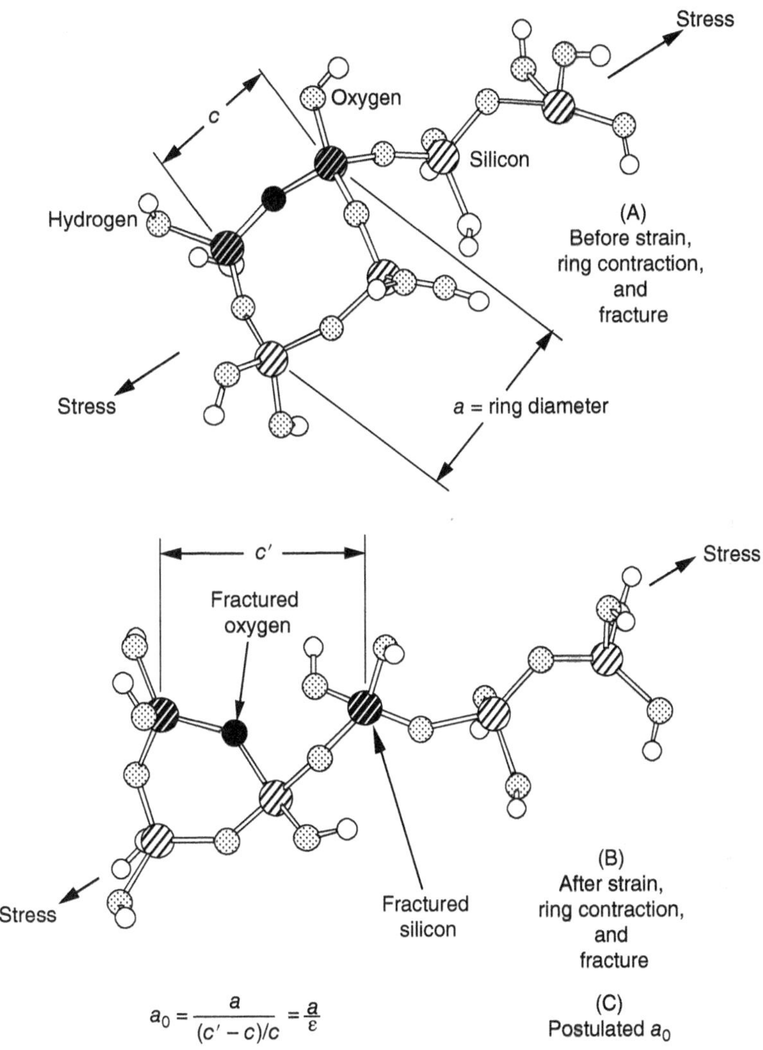

Figure 7.4 Schematic of bond fracture in silica. Source: West et al. (1999). Reproduced with permission of Elsevier.

the Richardson method (Richardson 1961; Hill et al. 2001). The fractal dimensions determined from simulations of the fractured surfaces over a single order of magnitude matched experimentally determined fractal dimensions for the same materials at length scales 1000–100 000 times larger as shown in Table 7.2 (Tsai et al. 1997). Thus, these successful

Table 7.2

Comparison of Experimental and MD Results for Fractal Dimension Values

Material fracture plane/surface	K_{IC} (MPa·m$^{1/2}$)	Fractal dimension (experimental)	Fractal dimension (MD simulation)
Si {100}/{110}	1.26±0.06	2.16±0.04	2.16±0.06
Si {110}/{100}	1.23±0.08	2.10±0.04	2.11±0.05
Si {111}/{110}	1.17±0.08	2.06±0.02	2.09±0.04
Silica (amorphous)	0.75	2.11 ± 0.02	2.1

Source: Tsai et al. (1997). Reproduced with permission of John Wiley & Sons.

comparisons provided confidence that the modeling was approximately correct. However, the actual mechanism and energetics of bond rupture were still unknown because of the limitations of MD models, i.e. ignoring the electronic structure.

Figure 7.4 shows a schematic of the atomic representation of the change in configuration of silica before and after fracture. The initial shape, which represents the structure with strain but before the fracture event, is characterized by "*a*," the body diagonal length, and "*c*," the initial length between the reference silicon atoms. After fracture, c deforms to c' so that the strain, ε, is $(c' - c)/c$. The ratio of a/ε has been identified as a_0, a structure constant related to the fractal nature of the surface formed (West et al. 1999). The quantity a_0 is proposed to be a spectral average of all reconfiguration events occurring during fracture. The comparison of the values of a_0 calculated from the MO model and from experiment, i.e. using Eq. (7.5), is shown in Figure 7.5 (West et al. 1999). Notice that there is reasonably good agreement considering the difference in length scales and in the nature of the calculations of the two estimates. A more specific interpretation of a_0 is discussed below.

In order to understand the meaning of a_0 and D^*, we can relate the above results to the work of Freiman and Mecholsky (2010) mentioned earlier and summarized in Eqs. (7.8) and (7.9) and presented here in a slightly different form:

$$\gamma = k'\varepsilon^2 E d_0 \tag{7.11}$$

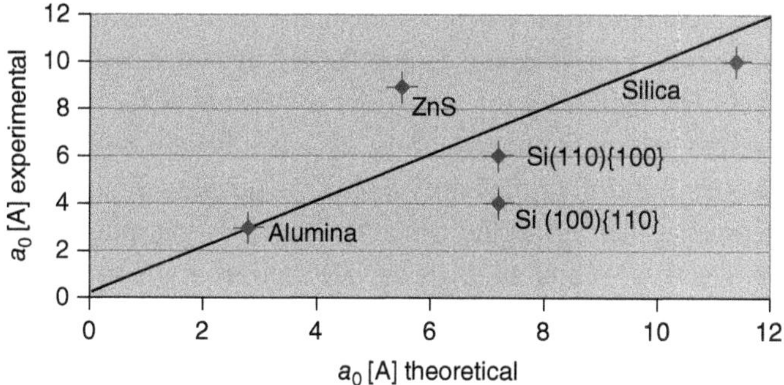

Figure 7.5 Experimental determination of a_0 using Eq. (7.10) versus the quantum mechanics calculation of a_0.

where k' is a constant dependent on the particular force function selected; E is the elastic modulus; d_0 is a structure parameter, usually the lattice spacing; and ε is the maximum strain at fracture. Notice the similarity between Eqs. (7.10) and (7.11). Equating Eqs. (7.10) and (7.11), since the values of γ must be the same, we obtain

$$a_0 ED^* = 2k'\varepsilon^2 E d_0 \tag{7.12}$$

Recall from the MO calculations that $a_0 = a/\varepsilon$, which is the ring diameter over the strain at fracture (West et al. 1999). Rearranging Eq. (7.12) and making the appropriate substitutions, Eq. (7.13) is obtained:

$$D^* = \left[\frac{2k'd}{a}\right]\varepsilon^3 = \alpha\varepsilon^3 \tag{7.13}$$

where α is another constant dependent on k', d_0, and a_0. Notice that the expression in the brackets contains atomic lengths and bonding parameters specific to each material. This result was interpreted to mean that the term ε^3 represents the amount of volume created during the bond reconfiguration event at one point along the crack front (Mečholsky et al. 2002). This reconfiguration process is one of many that occurs along the crack front during fracture. It was suggested by West et al. (1999) that this reconfiguration is the beginning of the fracture process.

Equation (7.13) also provides an insight as to the meaning of fractal scaling during fracture. The creation of the free volume caused by the instant reconfiguration when a critical energy level is reached during the fracture process provides the geometric step that creates part of the fracture surface. The parameter $<a_0>$ provides a measure of the structure upon which the free volume is created. Thus, $<a_0>$ is a linear representation of a volume element. It is not necessarily a physical quantity. In the language of fractal geometry, then D^* is the generator and a_0 is the initiator. Equation (7.13) implies that volume units of $\alpha\varepsilon^3$ are created at the tip of the crack, where α is a constant dependent on the structure, and in some cases grow at multiple units of $\alpha\varepsilon^3$ and in other cases are annihilated. In the cases where these regions grow, they eventually become the formation of mist, and then hackle, features observed on the fracture surfaces of materials that failed in a brittle manner. Assuming there was enough energy supplied at the beginning of fracture, the dissipation of this energy continues until the propagating crack cannot form enough surface to relieve the stored energy and the crack branches macroscopically. Macroscopic branching is also a fractal process but obeys different scaling rules than the formation of the surfaces (Mecholsky et al. 1989). The formation of the mirror–mist, mist–hackle, and macroscopic crack branching occurs at discrete energy levels (Tsai and Mecholsky 1991). However, the reason for the existence of these levels is still unknown. It is thought that with the understanding of the fractal nature of the fracture process, more insight into the reason for the required energy levels for microscopic and macroscopic branching will be possible.

Quantum mechanical methods are making great strides to understanding the required energy levels for both microscopic and macroscopic crack branching (Bartlett 2002; Musial and Bartlett 2011), but these methods are not easily accessible to materials science due to computational cost. The more affordable MBPT2, while ab initio, at times lacks the necessary quality. Strasberg (2013) has proposed an improved version of MBPT2 that should lead to a more effective approach to study fracture.

In any computational approach the final evaluation should be based on whether the model helps understand the fracture process such that accurate predictions of future behavior can be made. The most effective approach to this end is to use both computational and experimental methods in a complementary fashion. Experiments should be devised that can be tested with existing or newly developed computational models and vice versa.

QUESTIONS

1. What is the significance of Eq. (7.9) in terms of selecting a potential energy function to use in Eq. (7.2) for determining the fracture energy of single crystals?
2. What is the relationship between fracture energy and elastic modulus for single crystals? Does this relationship exist even for different crystal structures? Do you expect this same relationship to hold for polycrystalline materials? Why or why not?
3. Explain how the bond angle, number of bonding electrons, and total number of electrons would affect the strain of single crystals deformed in uniaxial tension.
4. Describe the West-Hench model of fracture in silica in a water environment.
5. Explain the similarities and differences between Eqs. (7.9) and (7.10). (Hint: Examine Eq. (7.13).)
6. How does the existence of fractal fracture affect the approach to modeling fracture in materials that fail in a brittle manner?
7. How would you select between molecular dynamics, molecular orbital, or quantum mechanics models in determining the fracture process in single crystals?

REFERENCES

Bartlett, R.J. (2002). To multireference or not to multireference: that is the question? *Int. J. Mol. Sci.* 3: 579–603.

Becher, P.F. and Freiman, S.W. (1978). Crack propagation in alkaline-earth fluorides. *J. Appl. Phys.* 49: 3779–3783.

Bell, R.J. and Dean, P. (1972). The structure of vitreous silica: validity of the random network theory. *Philos. Mag.* 25: 1381–1398.

Bitzek, E., Kermode, J.R., and Gumbsch, P. (2015). Atomistic aspects of fracture. *Int. J. Fract.* 191: 13–30.

Bouchaud, E., Lapasset, G., and Planès, J. (1990). Fractal dimension of fractured surfaces: a universal value? *Europhys. Lett.* 13: 73.

Bouchaud, J.P., Bouchaud, E., Lapasset, G., and Planès, J. (1993). Model of fractal cracks. *Phys. Rev. B* 48: 2917.

Chen, C.P. and Leipold, M.H. (1980). Fracture toughness of silicon. *Am. Ceram. Soc. Bull.* 59: 469–472.

Del Bene, J.E., Runge, K., and Bartlett, R.J. (2003). A quantum chemical mechanism for the water-initiated decomposition of silica. *Comput. Mater. Sci.* 27: 102–108.

Freiman, S.W. and Mecholsky, J.J. Jr. (2010). The fracture energy of brittle crystals. *J. Mater. Sci.* 45: 4063–4066.

Freiman, S.W., Becher, P.F., and Klein, P.H. (1975). Initiation of crack propagation in KCl. *Philos. Mag.* 31: 829.

Gilman, J.J. (1960). Direct measurements of the surface energies of crystals. *J. Appl. Phys.* 31 (12): 2208–2218.

Gumbsch, P. and Cannon, R.M. (2000). Atomistic aspects of brittle fracture. *MRS Bull.* 25 (5): 15–20.

Gutshall, P.L. and Gross, G.E. (1965). Cleavage surface energy of NaCl and MgO in vacuum. *Journal of Applied Physics* 36: 2459.

Hill, T.J., Della Bona, A., and Mecholsky, J.J. Jr. (2001). Establishing a protocol for optical measurements of fractal dimensions in brittle materials. *J. Mater. Sci.* 36: 2651.

Jaccodine, R.J. (1963). Surface energy of germanium and silicon. *J. Electrochem. Soc.* 110 (6): 524–527.

Kennedy, C.R., Bradt, R.C., and Rindone, G.E. (1973). Fracture mechanics of binary sodium silicate glasses. In: *Fracture Mechanics of Ceramics*, 2e (ed. R.C. Bradt, A.G. Evans, D.P.H. Hasselman and F.F. Lange), 883–893. New York: Plenum Press.

King, S.W. and Antonelli, G.A. (2007). Simple bond energy approach for non-destructive measurements of the fracture toughness of brittle materials. *Thin Solid Films* 515: 7232–7241.

Lawn, B.R. (1975). An atomistic model of kinetic crack growth in brittle solids. *J. Mater. Sci.* 10: 469–480.

Lazar, P. and Podloucky, R. (2008). Cleavage fracture of a crystal: density functional theory calculations based on a model which includes structural relaxations. *Phys. Rev. B* 78: 1, 104114–8.

Lazar, P., Podloucky, R., and Wolf, W. (2005). Correlating elasticity and cleavage. *Appl. Phys. Lett.* 87: 1, 261910–3.

Lindsay, C.G., White, G.S., Freiman, S.W., and Wong-Ng, W. (1994). Molecular-orbital study of an environmentally enhanced crack-growth process in silica. *J. Am. Ceram. Soc.* 77 (8): 2179–2187.

Mandelbrot, B.B., Passoja, D.E., and Paullay, J. (1984). Fractal character of fracture surfaces of metals. *Nature* 308: 721.

Margevicius, R.W. and Gumbsch, P. (1998). Influence of crack propagation direction on {110} fracture toughness of gallium arsenide. *Philos. Mag. A* 78 (3): 567–581.

Mecholsky, J.J., Freiman, S.W., and Rice, R.W. (1976). Fracture surface analysis of ceramics. *J. Mater. Sci.* 11: 1310–1319.

Mecholsky, J.J., Passoja, D.E., and Feinberg-Ringel, K.S. (1989). Quantitative analysis of brittle fracture surfaces using fractal geometry. *J. Am. Ceram. Soc.* 72 (1): 60–65.

Mecholsky, J.J. Jr., West, J.K., and Passoja, D.E. (2002). Fractal dimension as a characterization of free volume created during fracture in brittle materials. *Philos. Mag. A* 82 (17–18): 3163–3176.

Michalske, T.A. and Freiman, S.W. (1983). A molecular mechanism for stress corrosion in vitreous silica. *J. Am. Ceram. Soc.* 66: 284–288.

Muralidharan, K., Simmons, J.H., Deymier, P.A., and Runge, K. (2005). Molecular dynamics studies of brittle fracture in vitreous silica: review and recent progress. *J. Noncryst. Solids* 351 (18): 1532–1542.

Musial, M. and Bartlett, R.J. (2011). Charge-transfer separability and size-extensivity in the equation-of-motion coupled cluster method: EOM-CCx. *J. Chem. Phys.* 134: 034106-1–034106-12.

Novikov, N.V. and Dub, S.N. (1991). Fracture Toughness of Diamond Single Crystals. *J. Hard Mater.* 2: 3–11.

Raynes, A.S., Freiman, S.W., Gayle, F.W., and Kaiser, D.L. (1991). Fracture toughness of $YBa_2Cu_3O_{6+\delta}$ single crystals: anisotropy and twinning effects. *J. Appl. Phys.* 70 (10): 5254–5257.

Richardson, L.F. (1961). The problem of contiguity: an appendix of statistics of deadly quarrels. *Gen. Syst. Yearbook* 6: 139.

Rose, J.H., Ferrante, J., and Smith, J.R. (1981). Universal binding energy curves for metals and bimetallic interfaces. *Phys. Rev. Lett.* 47: 675–678.

Russ, J.C. (1994). *Fractal Surfaces*. New York: Plenum Press.

Shockey, D.A. and Groves, G.W. (1968). Effect of water on toughness of MgO crystals. *J. Am. Cer. Soc.* 51: 299–303.

Sih, G.C., Paris, P.C., and Irwin, G.R. (1965). On cracks in rectilinearly anisotropic bodies. *Int. J. Fract. Mech.* 1: 189–203.

Simmons, G. and Wang, H. (1971). *Single Crystal Elastic Constants and Calculated Aggregate Properties: A Handbook*, 2e. Cambridge, MA: MIT Press.

Stewart, R.L. and Bradt, R.C. (1980). Fracture of single crystal $MgAl_2O_4$. *J. Mater. Sci.* 15: 67–72.

Strasberg, M. (2013). Development of new quantum methods for materials science. PhD dissertation, University of Florida.

Swiler, T.P., Varghese, T., and Simmons, J.H. (1995). Evidence of chaotic behavior in computer simulations of fracture processes. *J. Noncryst. Solids* 181: 238.

Talman, J.D. and Shadwick, W.F. (1976). Optimized effective atomic central potential. *Phys. Rev. A* 14: 36–40.

Tang, Q.H. (2009). Effect of water on brittle fracture of $SiO_2$by molecular dynamics study. *Comp. Mat. Sci.* 45: 429–433.

Thomson, R., Hsieh, C., and Rana, V. (1971). Lattice trapping of fracture cracks. *J. Appl. Phys.* 42 (8): 3154–3160.

Tromans, D. and Meech, J.A. (2002). Fracture toughness and surface energies of mineral: theoretical estimates for oxides, sulphides, silicates and halides. *Miner. Eng.* 15: 1027–1041.

Tromans, D. and Meech, J.A. (2004). Fracture toughness and surface energies of covalent minerals: theoretical estimates. *Miner. Eng.* 17: 1–15.

Tsai, Y.L. (1993). Experimental and molecular dynamics determination of fractal fracture in single crystal silicon. PhD dissertation, University of Florida.

Tsai, Y.L. and Mecholsky, J.J. Jr. (1991). Fractal fracture in single crystal silicon. *J. Mater. Res.* 6 (6): 1248–1263.

Tsai, Y.L., Swiler, T.P., Simmons, J.H., and Mecholsky, Jr., J. J., (1997) "Fractal analysis of glass and crystals using molecular dynamics" in *Computational Modelling of Materials and Processing*, J. H. Simmons (ed. E.R. Fuller, A.L. Dragoo, and E.J. Garboczi) Ceramic Transactions 69 217. The American Ceramic Society, Westerville, OH

West, J.K. and Hench, L.L. (1994a). Silica fracture, part I: a ring contraction model. *J. Mater. Sci.* 29: 3601–3606.

West, J.K. and Hench, L.L. (1994b). Silica fracture, 2: a ring opening model via hydrolysis. *J. Mater. Sci.* 29: 5808–5816.

West, J.K. and Hench, L.L. (1995). Silica fracture, 3: five- and six-fold contraction models. *J. Mater. Sci.* 30: 6281–6287.

West, J.K., Mecholsky, J.J. Jr., and Hench, L.L. (1999). The application of fractal and quantum geometry to brittle fracture. *J. Noncryst. Solids* 260 (1–2): 99–108.

Westwood, A.R.C. and Goldheim, D.L. (1963). Cleavage surface energy of {100} magnesium oxide. *Journal of Applied Physics* 34: 3335.

Westwood, A.R.C. and Hitch, T.T. (1963). Surface energy of {100} potassium chloride. *J. Appl. Phys.* 34: 3085.

White, G.S., Freiman, S.W., Fuller, E.R. Jr., and Baker, T.L. (1988). Effects of crystal bonding on brittle fracture. *J. Mater. Res.* 2: 491–497.

Wong-Ng, W., White, G.S., and Freiman, S.W. (1992). Application of molecular orbital calculations to fracture mechanics: effect of applied strain on charge distribution in silica. *J. Am. Ceram. Soc.* 75 (11): 3097–3102.

Wong-Ng, W., White, G.S., Freiman, S.W., and Lindsay, C.G. (1996). Calculated potential for water enhanced crack growth in silicon. *Comput. Mater. Sci.* 6: 63–70.

Yuan, C.C. and Xi, X.K. (2011). On the correlation of Young's modulus and the fracture strength of metallic glasses. *J. Appl. Phys.* 109: 033515.

Zhu, T., Li, J., Lin, X., and Yip, S. (2005). Stress-dependent molecular pathways of silica-water reaction. *J. Mech. Phys. Solids* 53: 1597–1623.

CHAPTER 8

Quantitative Fractography

INTRODUCTION

When a material fractures in a brittle manner, there are characteristic markings left on the fracture surfaces that reveal the entire history of the fracture process. From these markings we can determine the fracture stress, identify the origin of fracture, establish the possible existence of local or global residual stresses, determine whether slow crack growth has taken place, and know the direction of crack propagation throughout the entire body. It is important to learn the techniques by which we can glean this information. Quantitative fractography has been used for decades to analyze the failure of industrial and laboratory components (De Freminville 1914; Preston 1926; Wallner 1939; Shand 1959; Rice 1984). In addition, the principles of fracture mechanics have been shown to be very powerful tools that complement the observations during fractographic analysis (Freiman et al. 1991).

The Fracture of Brittle Materials: Testing and Analysis, Second Edition.
Stephen W. Freiman and John J. Mecholsky, Jr.

FRACTURE MECHANICS BACKGROUND

Chapter 2 provides a background to fracture mechanics, some of which will be repeated here. Almost all of the mechanically induced cracks can be idealized as semielliptical, sharp cracks of semiminor axis, a, and semimajor axis, b. The relationship between the stress at fracture, or strength, σ_f, and the mode I stress intensity factor, K_I, is

$$K_I = \sigma M \left(\frac{\pi a}{\varphi^2}\right)^{1/2} \tag{8.1}$$

where ϕ is an elliptical integral of the second kind that accounts for the fact that the stress intensity factor varies around the perimeter of the flaw and is given by (Sanford 2003; Anderson 2005)

$$\varphi = \int_0^{\pi/2} \left[\sin^2\theta + \frac{a^2}{b^2}\cos^2\theta\right]^{1/2} \mathrm{d}\theta \tag{8.2}$$

where θ is the angle from the (tensile) surface toward the (interior) center along the crack periphery.

The stress intensity is a maximum at the tip of the minor axis of the flaw and is a minimum at the tip of the major axis. M is a surface correction factor whose value is approximately 1.12 as determined by finite element analyses (Raju and Newman 1979). The above relationship (Eq. 8.1) can be expressed in a simpler, useful manner for the critical condition of fracture as

$$K_{IC} = Y\sigma_f\left(c^{1/2}\right) = \sqrt{2E'\gamma_c} \tag{8.3}$$

where K_{IC} is the critical stress intensity factor (fracture toughness) and Y is a geometric factor that accounts for the shape and location of, and loading on, the fracture initiating crack; E' is E, the elastic (Young's) modulus for plane stress conditions, and $E' = E/(1-\nu^2)$ for plane strain conditions where ν is Poisson's ratio; and γ_c is the critical fracture energy, i.e. all the energy involved in fracture including the creation of new surface (Bar-on 2001). The quantity Y in Eq. (8.3) depends on the ratio a/b as described by ϕ in Eqs. (8.1) and (8.2). However, the crack size can be approximated using an equivalent semicircular crack size, c

[$c = (a * b)^{1/2}$]. This approximation allows many irregular crack shapes to be analyzed and avoids the complications of calculating a geometric factor for each crack (Randall 1967; Mecholsky et al. 1977b). For surface cracks without local residual stress and those that are small relative to the thickness of the sample, $Y \sim 1.26$.

FRACTURE SURFACE OBSERVATIONS

Sources of Failure

Fracture in brittle materials occurs due to the application of stress to a flaw in the material such that the combination of stress and size of flaw leads to unstable propagation of that defect. In this context, flaw, crack, and fracture origin are used interchangeably. Almost all materials have "flaws" due to fabrication, manufacturing, or preparation (see, e.g. Choi and Gyekenyesi 1998; Danzer et al. 2008). Most fracture origins in brittle materials are cracks that tend to be effectively two-dimensional, i.e. their opening is of the order of nanometers. Even when the source is an inclusion of foreign material, the fracture generally originates from a crack at the boundary between the bulk material and the foreign material, e.g. observe Figure 8.1. Pores can act as a source of failure and are not as severe as a sharp crack. When pores act as a source of failure, they are much larger than any sharp crack in the same stressed region (Rice 1994). As an example of a large pore not effective enough to be an origin, see Figure 8.2.

The shape and size of cracks are important to the fracture process, so it is necessary to understand and to investigate their sources and characterization. The types of sources can be categorized by their origination: extrinsic and intrinsic. Extrinsic origins include mechanical (handling, intentional and unintentional indentation, finishing operations, impact damage, etc.) (Mecholsky et al. 1977b) and chemical and thermal processing and treatment (sputtering, etching, quenching, thermal shock, etc.) (e.g. cf. Danzer et al. 2008). Intrinsic origins include defects in the structure inherent in the specific fabrication or manufacturing technique such as sintering that can lead to low density regions or extrusion that can lead to nonuniform grains (Rice 1984; Danzer et al. 2008).

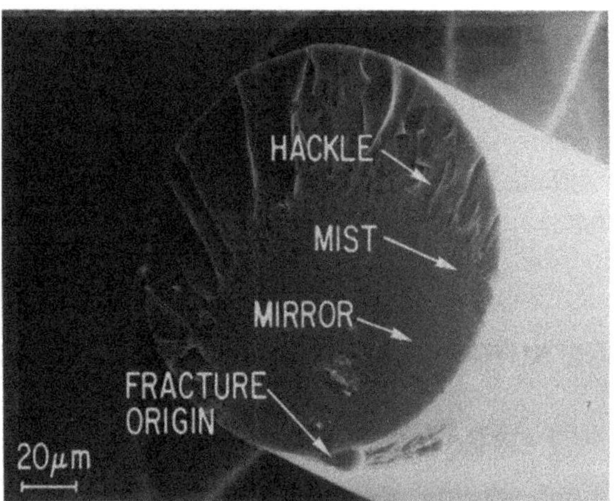

Figure 8.1 Fracture surface of a silica glass fiber showing fracture origin, mirror, mist, and hackle regions. The origin is a crack at the surface created by the thermal expansion difference between the fiber and the foreign "dust" particle attached to it while the fiber was still hot during fabrication. The actual fracture origin is a crack on the glass surface at the dust particle.

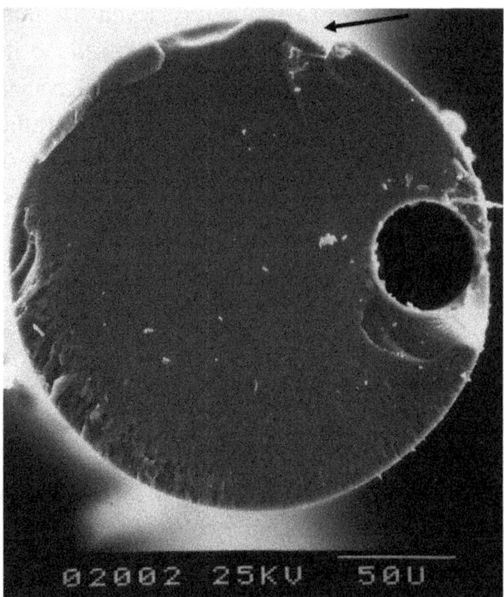

Figure 8.2 Optical fiber fractured in bending. Arrow points to the origin. Notice that the pore or bubble is well away from the origin of failure even though it is much larger. In this particular case, the stress was not great enough in the region of the hole to cause a large enough stress concentration.

Fracture origins may be characterized by their identity, location, and size. Fracture origins may be volume flaws such as internal cracks, pores, agglomerates, inhomogeneous density, or compositional regions or surface origins such as cracks due to machining, surface pits or voids, and impact damage due to handling (Mecholsky 1992; Quinn and Swab 1996). Of course, a volume-type defect can be located at the surface, and a material can contain both surface and volume flaws. The location can be at the surface, in the volume, near the surface, and/or at the edge of a component (Danzer et al. 2008).

Propagating Cracks

As seen in Figures 8.3 and 8.4, there are five primary regions of crack growth observable on a fracture surface. These are (i) the flaw itself, (ii) the "mirror" region, (iii) the mist region, (iv) the hackle region, and (v) the region of macroscopic crack branching. Often the original crack leaves a discernable boundary such as that indicated in Figure 8.1. The crack grows catastrophically from the boundary at K_{IC} if there is no slow crack growth prior to rapid propagation. In the case of slow crack growth preceding rapid fracture, there may not be a discernable mark on the surface to mark the point at which the crack reaches K_{IC}. As the crack grows from the fracture origin with increasing velocity, the crack

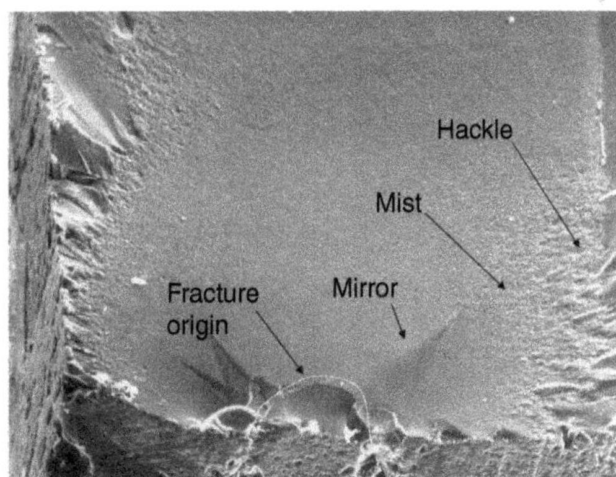

Figure 8.3 Scanning electron micrograph of the fracture surface of a glass rectangular bar broken in flexure. The white line is drawn just outside the critical crack that led to failure.

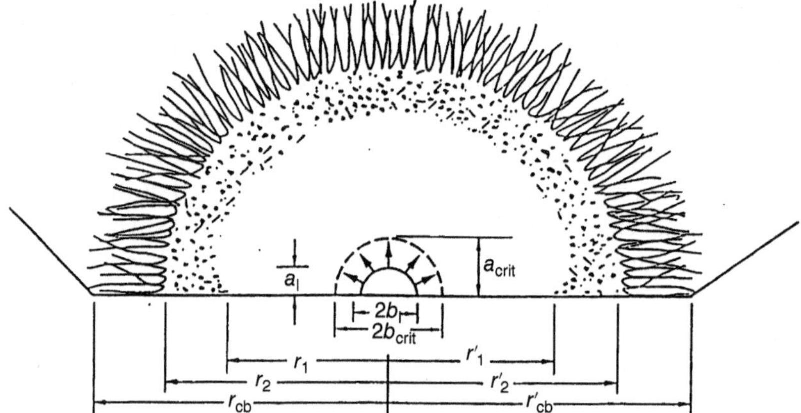

Figure 8.4 Schematic of fracture surface of material failed in a brittle manner. The critical crack (a_{crit}, b_{crit}) is indicated by the dashed line. The solid line (a_i, b_i) indicates the initial crack size. The radii, r_j, indicate the distance from the origin to the respective boundaries between the mirror–mist ($j = 1$), mist–hackle ($j = 2$), and macroscopic crack branching ($j = 3$ or cb). The prime indicates that the values on either side of the crack may be different.

first propagates in a relatively smooth plane (the mirror region) to the boundary, r_1, and progressively gets rougher by deviating slightly out of plane in a region that resembles mist (the mist region), between r_1 and r_2 (Figures 8.3 and 8.4). Finally, the crack deviates locally from the main plane of fracture (the hackle region), between r_2 and r_3, getting very rough and finally branching into two or more cracks at r_3. Beyond macroscopic crack branching, the process can repeat itself on each branch of the propagating crack (Rossmanith 1980; Quinn 1999, 2016). These regions are all related to the applied (far-field) stress at fracture, σ_f:

$$\sigma_f r_j^{1/2} = \text{constant} = A_j = \frac{K_{Bj}}{Y'(\theta)} = \sqrt{\frac{2E'\gamma_{Bj}}{Y'(\theta)}} \tag{8.4}$$

where $r_j = r_1$, r_2, or r_3 corresponding to the different regions in an analogous equation to Eq. (8.3) and A_j is the corresponding "mirror" constant (Kirchner and Conway 1988; Mecholsky 1991). K_{Bj} is the crack branching stress intensity and $Y'(\theta)$ is a crack border correction factor where θ is the angle from the (tensile) surface to the interior, i.e. $\theta = 0$–$90°$, analogous to the crack geometry (Kirchner and Kirchner 1979). Note that K_{Bj} is proportional to K_{IC}, i.e. $K_{Bj} = \lambda K_{IC}$ where λ is

2–4 (for $j = 3$) for most ceramic materials (Kirchner 1986; Choi and Gyekenyesi 1998). γ_{Bj} is the branching energy associated with the different regions, and E' is the same as above in Eq. (8.3) (Mecholsky 1991).

Figure 8.1 shows an optical fiber that failed in bending (Mecholsky et al. 1977a). Even though the fiber is relatively small, it still shows the characteristic features demonstrated in Figure 8.4. Figure 8.5 shows the strength of silica-based glass versus the fracture mirror radius, r_1, for fibers loaded in tension and bending, for disks loaded in biaxial tension, and for bars loaded in three-point flexure. The straight line represents Eq. (8.4) with $j = 1$. The fact that the relationship represented by Eq. (8.4) is valid for several stress states during failure provides increasing confidence that this equation is useful in analyzing *in situ* field failures. Also, notice the large range of stress values and of sizes of fracture mirror radii. For an estimate of the accuracy of these types of measurements, see Salem and Jenkins (2002). The values of the fracture toughness and crack branching constants for several materials are presented in Table 8.1.

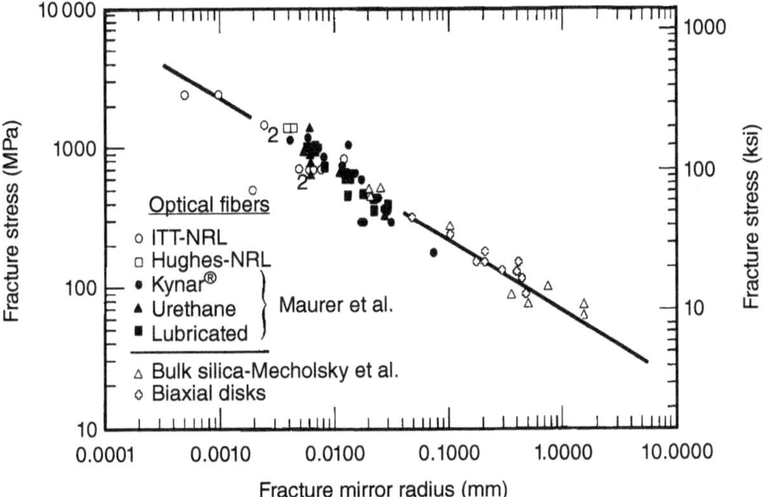

Figure 8.5 Graph of the fracture stress for silica-based glass as a function of the mirror–mist radius (r_1 in Eq. 8.4). The Maurer et al. (1974) data are fibers failed in tension. The ITT-NRL and Hughes-NRL data are also fractured in tension. The bulk silica data were rectangular flexure bars (Mecholsky et al. 1974).

Table 8.1
Fracture Properties of Brittle Materials

Material	Source	E (GPa)	γ (J m^{-2})	K_{IC} (MPa·m$^{1/2}$)	K_2 (MPa·m$^{1/2}$)	K_3 (MPa·m$^{1/2}$)
1. ADP [SC][a]	M[b]	9	2	0.2	0.6	
2. $MgAl_2O_4$ [SC]	M	240	13	0.7	3.22	
3. MgO [SC]	M	280	3	1.3	6.2	
4. Al_2O_3 [SC]	M	340	7	2.2	7.5	
5. AlSiMag 614 [SC]	M	300	20	3.5	16.2	
6. Al_2O_3 [SC]	M	210	13	2.3	8.0	
7. 96% Al_2O_3 (flex)	K	320	19.5	3.2		10
8. 95% Al_2O_3 (delayed)	M	320	19.5	3.2	10.3	
9. HP Al_2O_3 (flex)	M	380	34	5.1	11.9	
10. HP Al_2O_3 (delayed)	M	380	34	5.1	11.4	
11. 96% Al_2O_3	C			3.1		8.9
12. 96% Al_2O_3	C			3.1		9.2
13. 96% Al_2O_3	C			3.1	9.4	
14. B_4C	M	450	15	3.7	11.4	
15. $BaTiO_2$	M	120	5	1.1	6.2	
16. $BaTiO_2$ (LiF-MgO)	M	120	6	1.2	6.7	
17. CaF_2	M	114	0.5	0.3		
18. CdTe	M	40	1	0.3		
19. Glassy carbon	M	25	8.5	0.7	2.1	
20. Graphite	M	12	85	1.4	4.1	
21. HP Al_2O_3 (99+)	K	390	20	4.0		12.8
22. $MgAl_2O_4$	M	240	6	1.7	9.6	
23. MgF_2	M	110	4	0.9	3.8	
24. MgO	K	280	12	3		7.6
25. Mullite	M	222	11	2.2	7.5	
26. PZT	M	80	4	0.8	4.5	

Table 8.1

(Continued)

Material	Source	E (GPa)	γ ($J\,m^{-2}$)	K_{IC} ($MPa \cdot m^{1/2}$)	K_2 ($MPa \cdot m^{1/2}$)	K_3 ($MPa \cdot m^{1/2}$)
27. Si_3N_4 (B)	M	280	30	4.1	14.8	
28. Si_3N_4 (HS-130)	M	310	45	5.3	22.4	
29. HP Si_3N_4 NC-132 (flex)	M	310	39	4.7	10.3	
30. HP Si_3N_4 NC-132	C	310	39	4.7		13.0
31. HP Si_3N_4 (flex)	M	310		4.7	10.5	
32. HS-130	M	310		4.7	10.5	
33. RB Si_3N_4 (flex)	M	135	29	2.8	4.8	
34. HP SiC (flex)	C	440	18	4.1		16.7
35. HP SiC (delayed)	M	440	18	4.1	13.8	
36. $SrZrO_3$	M	280	6	1.8	7.4	
37. SiC (XT)	M	390	19	3.9	13.2	
38. Silicate glasses	M	64	3	0.7	2.6	
39. Spinel	M	240	3	1.2		
40. ZnSe	M	69	5.5	0.9		2.1
41. ZrO_2 (Zircar)	M	280	70	6.3	18.8	
42. ZrO_2 (zyttrite)	M	260	13	2.6	9.1	
43. Zircon porcelain	M	160			4.6	
44. Steatite (clinoen-statite)	K	110		2.7		5.6
45. $3BaO\text{-}5SiO_2$ [GC]	M	80	17	2.2	7.4	
46. Cervit 126 [GC]	M	92	17	1.8	7.3	
47. $Li_2O\text{-}2SiO_2$ [GC]	M	70	7	1.0	5.5	
48. $Li_2O\text{-}2SiO_2$ [GC]	M	137	33	3	6.7	
49. Lithia borosilicate [GC]	M	90	40.5	2.7		
50. Pyroceram 9606 [GC]	M	120	25	2.5	8.0	

(Continued)

Table 8.1
(Continued)

Material	Source	E (GPa)	γ (J m^{-2})	K_{IC} (MPa·m$^{1/2}$)	K_2 (MPa·m$^{1/2}$)	K_3 (MPa·m$^{1/2}$)
51. Flint glass [G]	K	69		0.7		2.6
52. Silicate glasses [G]	M	69	3	0.7	2.6	3.0
53. Soda-lime-silicate glass [G]	C			0.76		4.3
54. Fused silica fibers [G]	C			0.79	2.6	
55. Silica glass [G]	K	72	4	0.8		3.0

[a] SC, single crystal; GC, glass ceramic; G, inorganic glass. Others are polycrystalline materials.
[b] M, Mecholsky (2001); C, Choi and Gyekenyesi (1998); K, Kirchner (1986).

QUANTITATIVE ANALYSIS

In general, brittle fracture begins from one primary failure origin. The location of that origin is at the largest crack or flaw in the highest stressed region. The crack at the origin propagates approximately perpendicular to the principal tensile stress (alternatively, where K_{II} is zero) in an unstable manner and, for a uniform stress field, radiates out in an approximately symmetric manner from the origin (Figures 8.3 and 8.4) (Freiman et al. 1991). Due to thermal vibrations, a distribution of bond strengths, and an irregular crack front, the propagation of the crack does not occur on one plane (Beauchamp 1996). The crack immediately leaves the plane perpendicular to the applied stress direction at specific points along the crack front. Note that these departures from the plane initially have amplitudes smaller than the wavelength of light and would appear smooth under the optical microscope (Smith et al. 2009).

During loading, strain energy is stored in the material. The stored strain energy is released during propagation. There are many energy-absorbing processes that occur during crack propagation. Researchers have measured local increases in temperature at the crack tip and have observed the ejection of electrons, ions, and molecules during fracture (Kulawansa et al. 1994). When the creation of surface, the release of particles, the rise in temperature, and all other processes occurring during crack propagation are not great enough to satisfy the release of the stored strain energy, the main crack branches into two or more branches (Figure 8.4). If there is a supply of strain energy, the process will continue until there is no more material to fracture, or the energy is dissipated and the crack arrests.

The techniques associated with fractography and the identification of features are well established (Rice 1984; Frechette 1990; Mecholsky 1996; Morrell 2006; Quinn 2016). Excellent presentations of the principles of fractography can be found in the "Guide to Fractography of Glasses and Ceramics," a National Institute of Standards and Technology (NIST) publication (Quinn 2007), in ASTM Standard C1322-96a, and in the US Army Military handbook MIL-HDBK-790. Excellent correlation between Eq. (8.3) and the measurement of fracture initiating cracks in many glasses (Mecholsky et al. 1974) and ceramics (Mecholsky et al. 1976; Choi and Gyekenyesi 1998) is recorded in the literature. The fractographic principles apply to all classes of materials, i.e. ceramics, metals, or polymers. As an example, Figure 8.6a shows the fracture surface of single crystal silicon (Tsai and Mecholsky 1991), Figure 8.6b of single crystal alumina (Quispe-Cancapa et al. 2008), and Figure 8.7 of a large-grained ZnSe (Freiman et al. 1975). Single crystals fracture in a similar manner as polycrystalline materials except that the characteristic features on the fracture surface are affected by the anisotropy of the crystals. In addition, there is not usually a "mist" region. The boundary between the "mirror" region and the "hackle" region in single crystals, as well as isotopic materials, is a constant energy boundary (Tsai and Mecholsky 1992). In materials whose microstructure consists of large-grain materials, the features are often difficult to observe because of large local crack deflections in the direction of easy propagation of grain facets or boundaries (Rice et al. 1980). These large deflections mask the more subtle mist and hackle markings. Usually, this difficulty is compensated for because most large-grain materials fail in a transgranular manner and leave classic characteristic markings on the surface, e.g. cleavage marks, and twist hackle ("river" lines) indicators (Figure 8.7). The twist hackle (river) marks result from the propagating crack deflecting onto parallel planes in local, adjacent regions along the crack front. These regions also result in cleavage "steps" that are approximately perpendicular to

(a)

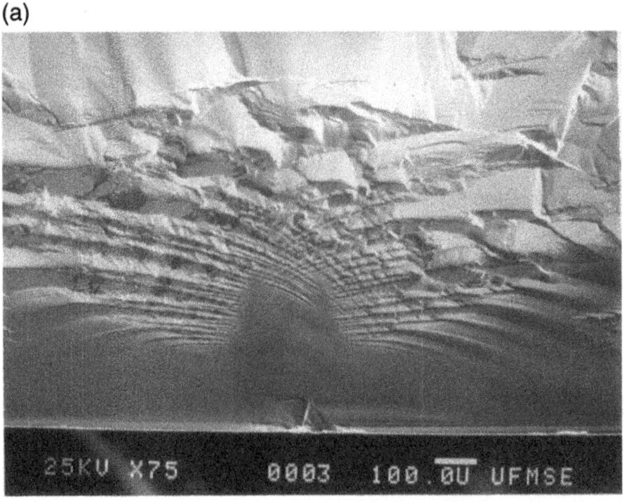

(b)

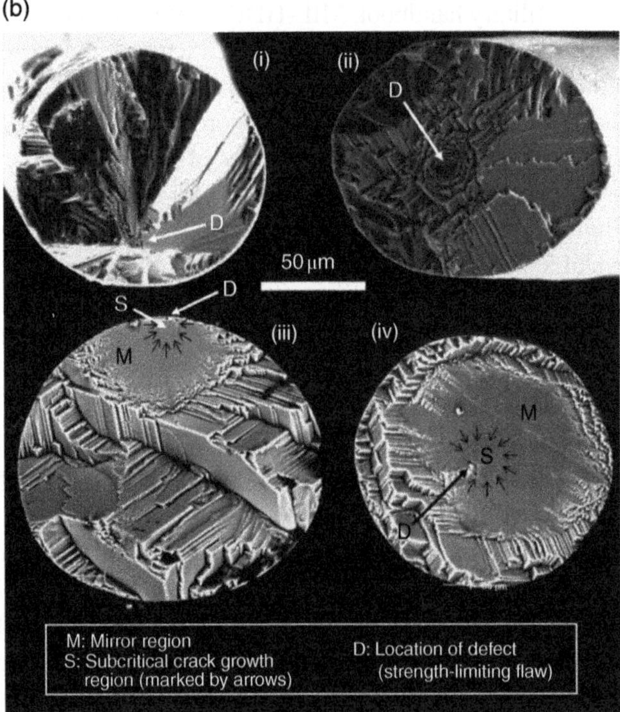

Figure 8.6 (a) Scanning electron micrograph of a single crystal silicon fracture surface. The features are different from the schematic of Figure 8.2 due to the anisotropic elastic moduli. However, the relationships shown in Eqs. (8.3) and (8.4) are still valid. (b) SEM images of single crystal alumina. (i) Pure sapphire, room temperature; (ii) Cr-doped sapphire, room temperature; (iii) pure sapphire, high temperature (1400 °C), with surface-originating defect; and (iv) Cr-doped sapphire, high temperature (1400 °C) with internal origin. Source: Quispe-Cancapa et al. (2008). Reproduced with permission of John Wiley & Sons.

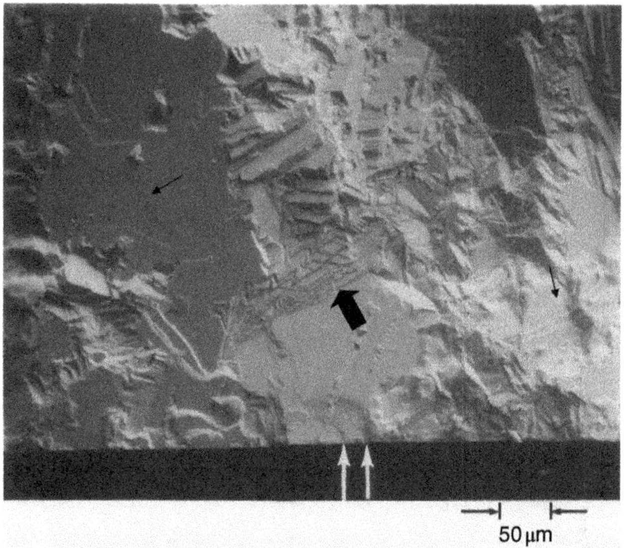

Figure 8.7 SEM image of a ZnSe fracture surface. This is a large-grain material that failed at a relatively low stress. The features in the image are all in the "mirror" region. However, Eq. (8.3) is applicable. The critical crack is the larger arc shown by the outer (white) arrow. The black arrows show twist hackle (river marks) that fan out from the origin and show direction of local crack propagation. Cleavage steps are shown with the large black arrow.

the crack propagation direction. Polycrystalline fracture surfaces such as shown in Figure 8.8 have the same characteristic features as stylized in Figure 8.4. However, the microstructure often obscures the identifying marks, and more experience is necessary to differentiate the regions. With the use of the atomic force microscope, it is possible to determine the regions in which the roughness increases in magnitude, thus delineating the transition regions between mirror, mist, and hackle.

The characteristic markings on the fracture surface can be used to determine the strength of a fractured part and to identify the cause of the failure. There is other information that can be gleaned from observing the fracture surface and measuring the crack size and mirror boundaries.

Global Residual Stress

Near-surface residual stresses that cover the entire outer surfaces of a material, σ_r, can greatly affect the performance of a part. If these global stresses are tensile ($+\sigma_r$), then the effective strength of the material is

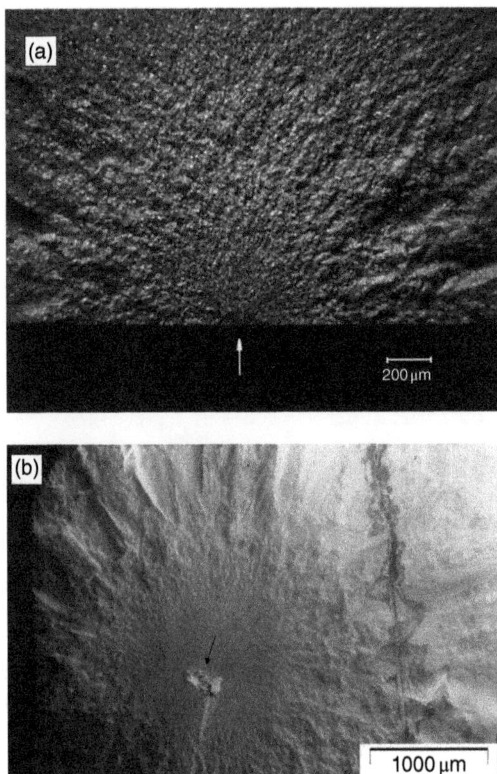

Figure 8.8 (a) SEM image of sintered reaction-bonded silicon nitride fractured with a crack (arrow) introduced by indenting with a 49 N Knoop diamond. Source: Quinn et al. (2003). Reproduced with permission of John Wiley & Sons. (b) SEM image of a silicon nitride fracture surface showing internal origin (arrow). This fracture is from a turbine rotor fractured in a spin test. Notice the relatively smooth region around the origin that transitions to rougher regions. Source: Danzer et al. (2008). Reproduced with permission of WILEY-VCH Verlag GmbH & Co.

reduced. If the global near-surface stresses are compressive ($-\sigma_r$), then this often leads to superior mechanical performance. Quantitative fractography can be used to evaluate the magnitude of these global residual stresses in postmortem analysis. Equation (8.4) can be modified to include the effect of residual near-surface stresses, σ_R (Conway and Mecholsky 1989):

$$\sigma_c r_j^{1/2} Y(\theta) = +K_{Bj} - \sigma_R r_j^{1/2} Y_R(\theta) \tag{8.5}$$

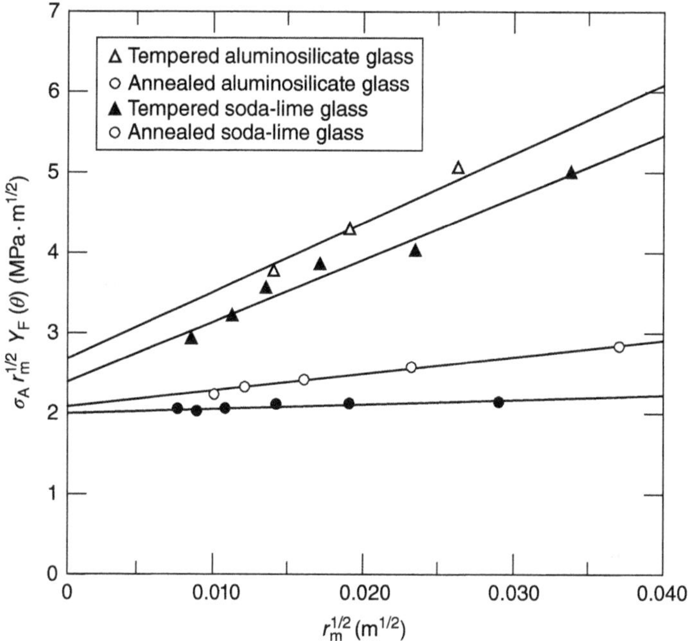

Figure 8.9 A graph of Eq. (8.5) with $r_m = r_1$. Source: The data are taken from the literature as described in Conway and Mecholsky (1989). Reproduced with permission of John Wiley & Sons.

where $Y_R(\theta)$ is a crack border correction factor for the residual stresses and usually assumed to be 1 and the other symbols are as described in Eq. (8.4). A graph of the left side of Eq. (8.5) versus a selected value of r_j, e.g. r_1, for several samples will result in data that have either a zero slope, i.e. no residual stress, or a slope related to σ_R (Figure 8.9). In this particular example, two compositions of glasses, aluminosilicate and soda-lime-silica, were annealed; some of these were tempered to produce global stresses, i.e. residual compressive stresses.

The annealed glasses had almost zero slope. (The fact that the aluminosilicate did not have zero slope implies that it was not fully annealed.) The tempered glasses had a slope related to the magnitude of the residual stress. For individual components, the expected applied stress for samples without residual stress can be estimated from Eq. (8.3) after measuring the crack size or from Eq. (8.4) after measuring the mirror radius. The mirror-to-flaw size ratio will be the same as without global residual stress. However, the shape of the mirror–mist border will be quite different. An example of this shape difference is illustrated in Figure 8.10a and b. Both fractures shown in Figure 8.10a occurred in flexure from the same-size flaw. The difference in shape is due to global residual surface compression in the lower figure of Figure 8.10a. Note that the distance of the mirror boundaries are the same on both surfaces, indicating that they both failed at the same stress level. However, the external loading was different, so the recorded stress at fracture would be greater for the material under residual compressive stress (lower figure). Since the crack size is the same and the mirror boundaries on the tensile surface are the same for both, then the mirror-to-flaw size ratio will be the same. In Figure 8.10b, the failure is from an indentation crack in a tempered glass, while the compressive stress was still present during fracture, as in the same case for the schematic (Figure 8.10a, lower figure).

Local Residual Stress

A typical crack may experience other components of residual stress, such as local residual stress produced by the creation of the crack from impact or contact events (Marshall and Lawn 1986). Local residual stress means that the effect of the stress is localized around the crack and effects of the residual stress disappear a few crack lengths away. Such is the case for deformation-induced surface cracks produced in sharp-point contact. The mirror-to-flaw size ratio for cracks with local residual stress is (Marshall et al. 1980)

$$\frac{r_j}{c} = \left[\frac{4YK_{Bj}}{3Y'(\theta)K_{IC}}\right]^2 \tag{8.6}$$

where Y is a constant that accounts for crack and loading geometry, $Y'(\theta)$ is the same type of constant as Y for the mirror boundaries, and the other terms are as defined previously. Along the tensile surface, we assume that Y and $Y'(\theta)$ will be equal. The value of r_j/c for cracks without local stress removes the factor $(4/3)^2$:

$$\frac{r_j}{c} = \left[\frac{YK_{Bj}}{Y'(\theta)K_{IC}}\right]^2. \tag{8.7}$$

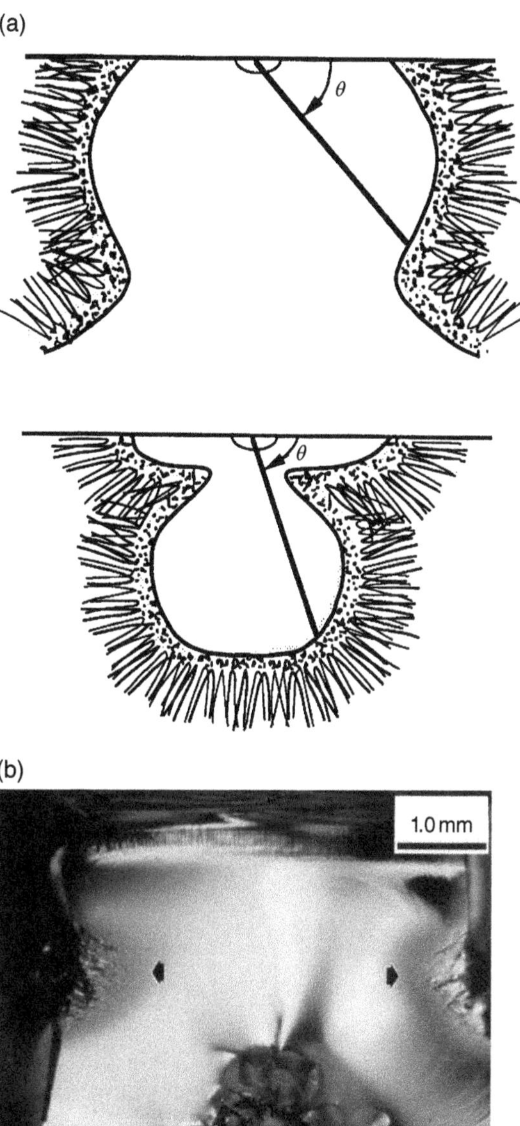

Figure 8.10 (a) Schematic of fracture surfaces with and without global residual stress. Both fractures occurred in flexure from the same-size critical crack. The difference in shape is due to residual surface compression in the lower figure. Note that the distance of the mirror boundaries are the same on both surfaces, indicating that they both failed at the same stress level. However, the external loading was different. (b) Fracture origin in tempered glass. Notice mirror–mist boundary (arrows) that mimic the schematic of (a). Source: From Beauchamp and Matalucci (1998).

Thus, when local residual stress is present, r_j/c is a constant that is 1.78 [$(4/3)^2$] times the r_j/c value when no residual stress is present. This also means that $Y = 1.68$ in Eq. (8.3) [$4/3 \cdot 1.26$] when local residual stress is present due to indentation or impact.

Slow Crack Growth

As discussed in Chapter 3, environmentally enhanced crack growth frequently occurs in ceramics, leading to delayed failure under stress. Fractographic analysis can be used to estimate the time to failure in delayed failure environments under constant load (Mecholsky et al. 1979; Michalske 1979). Assuming that the crack growth follows a power law-type behavior

$$V = V_0\left(\frac{K_{\mathrm{I}}}{K_{\mathrm{IC}}}\right)^N = AK_{\mathrm{I}}^N \tag{8.8}$$

where V is the crack velocity in region I of crack growth (cf. Chapter 3), V_0, A, and N are experimentally determined parameters for a material, and K_0 is a normalizing factor usually taken as $1\,\mathrm{MPa}\cdot\mathrm{m}^{1/2}$, then it can be shown that the time to failure, t_{f}, can be related to the *initial* crack size, c_{i}, the mirror–mist boundary, r_1, and "mirror" constant, M_1:

$$\frac{t_{\mathrm{f}}}{c_{\mathrm{i}}} = \frac{(1.24)^n (r_1/c_{\mathrm{i}})^{n/2}}{M_1^n A[(n/2)-1]} \tag{8.9}$$

Equation (8.9) is graphed in Figure 8.11 for equivalent semicircular cracks in soda-lime-silica glass bars. The dashed line is a least squares fit to the data ($r^2 = 0.73$). The dash-dot line represents the r_1/c_{crit} value of 11.6 ± 1.2 and is drawn parallel to the ordinate as an aid to the eye. The variable c_{crit} is the critical crack size calculated using Eq. (8.3). We can conclude from this analysis that when slow crack growth occurs, the ratio of the fracture mirror radii to initial crack size is a function of time, whereas the fracture mirror radius to critical crack size is a constant. The fracture mirror constants determined in rapid and in delayed failure are identical because, once the

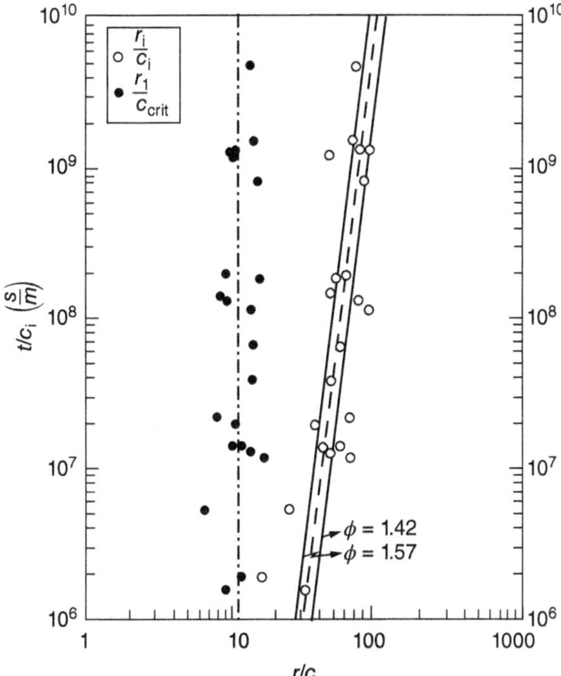

Figure 8.11 Time to failure as a function of the mirror-to-crack size ratio for soda-lime-silica glass. The value of $\phi = 1.57$ is for a semicircular crack (Eq. 8.1) and of $\phi = 1.42$ is for $a/b = 0.8$. The dot-dash line is drawn parallel to the ordinate and has a value of $r_1/c = 11.6 \pm 1.2$ for fast fracture. Source: Mecholsky et al. (1979). Reproduced with permission of John Wiley & Sons.

critical crack size is reached, the processes leading to rapid crack propagation and mirror formation are the same. At the time of rapid failure, the mirror radius-to-critical crack size ratio is the same as for the condition of annealed, stress-free material. However, the mirror-to-initial crack size is a function of time under load. Thus, if the mirror-to-crack size ratio is greater than about 20 : 1, then one may conclude that some slow crack growth occurred before fracture (cf. Figure 8.12). One can then also estimate the amount of time under load by using Eq. (8.9). Of course, if the stress was not constant during the loading, then modifications need to be made in the equation for variable stress and time under load.

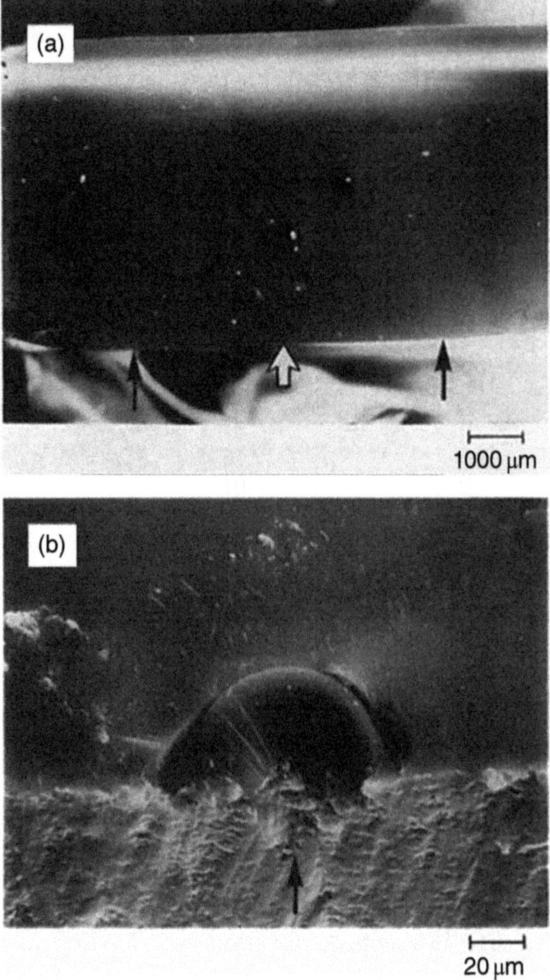

Figure 8.12 SEM image of fracture surface of a soda-lime-silica glass after delayed failure. Notice that the mirror boundaries (black arrows in (a)) are over 20 times greater than the initial crack size (white arrow in (a)) [$r_1/c \cong 80$]. Initial crack size from machining is shown in (b).

FRACTOGRAPHIC PROCEDURES

Fracture origins can be found by examining the fracture surface and tracing the characteristic markings back to the point from which they emanated. The fracture progresses from the original crack and expands

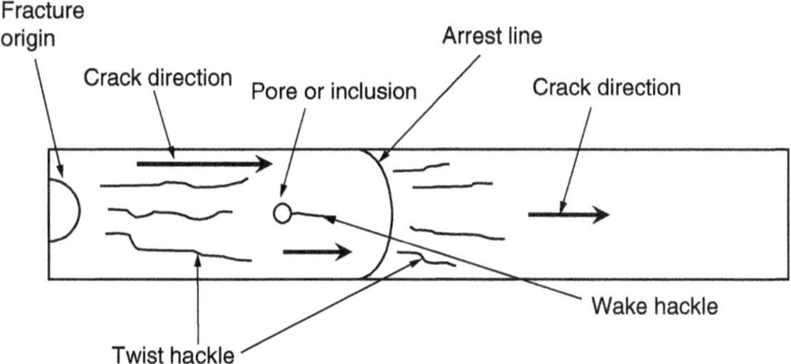

Figure 8.13 Schematic of fracture surface showing possible features on the surface to aid in determining origin and crack direction. Source: Courtesy of E.K. Beauchamp.

outward almost in a plane. A schematic representation of the possible features appearing on the fracture surface is presented in Figure 8.13. Typically, the fracture surface should be first examined with the aided eye, e.g. using a magnifying glass. The next level of observation should be with an optical microscope with several magnifying lenses from 50× to 200×. Most quantitative measurements should be made using the optical microscope. Typically, both low magnification and high magnification levels are used alternately in one examination. This procedure is used because features that show direction of propagation may be quite small; however, the critical crack may be much larger, and for the entire crack to be observed, a lower magnification may be needed. The scanning electron microscope (SEM) is very valuable for emphasizing certain fracture surface features, especially those with varying depths of field. In general, the SEM should not be used for quantitative measurements. A very excellent guide to the observation process is provided in the NIST Recommended Practice Guide (Quinn 2007). An example of fracture in a polycrystalline material demonstrating some of the features that are typically observed (Taskonak et al. 2008) is shown in Figure 8.14. These characteristic markings include twist hackle (river marks) (Figure 8.7), wake hackle (fracture tails) (Figure 8.15), cleavage steps (Figure 8.7), Wallner lines (Figure 8.16), arrest lines (Figure 8.16), and branching locations (Figure 8.17). Wallner lines are the result of intersections of the crack front with a propagating elastic stress wave initiated from another source, such as the surface, an inclusion, or

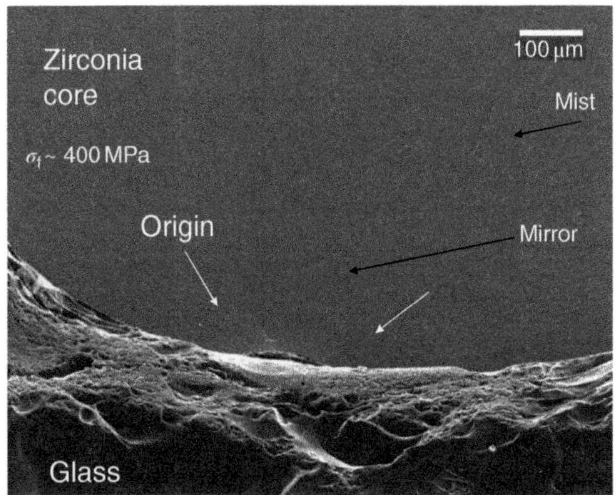

Figure 8.14 SEM image of a dental crown with zirconia core and a glass veneer. The secondary failure occurred at the interface between the glass and core. The critical crack is shown by the white arrows. The mirror and mist regions are shown but are more difficult to discern in polycrystalline materials. Source: Taskonak et al. (2008). Reproduced with permission of Elsevier.

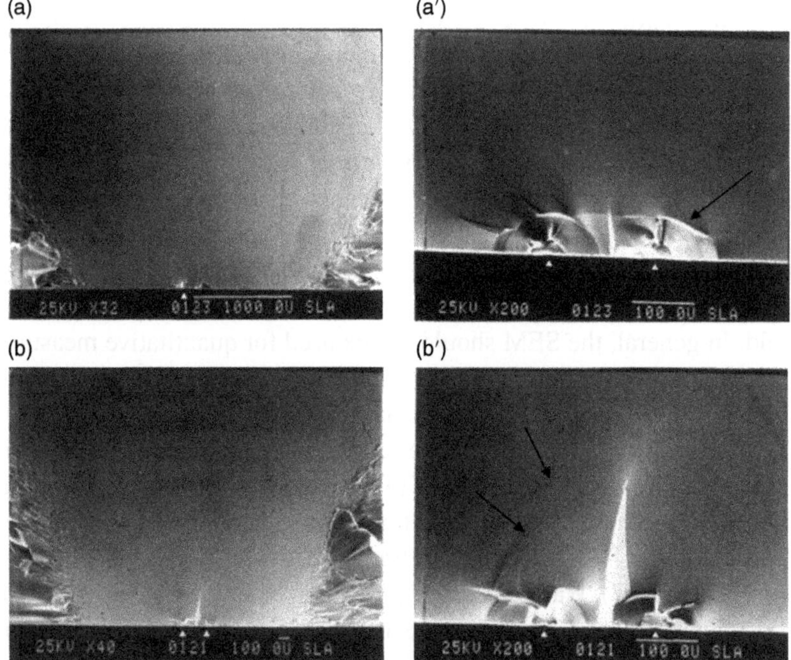

Figure 8.15 SEM images of soda-lime-silica glass fractured in flexure from indentation cracks. (a′) and (b′) show greater magnifications of (a) and (b). In (b′), the arrows point to Wallner lines. In (a′) the arrow points to arrest lines at the edge of the critical crack.

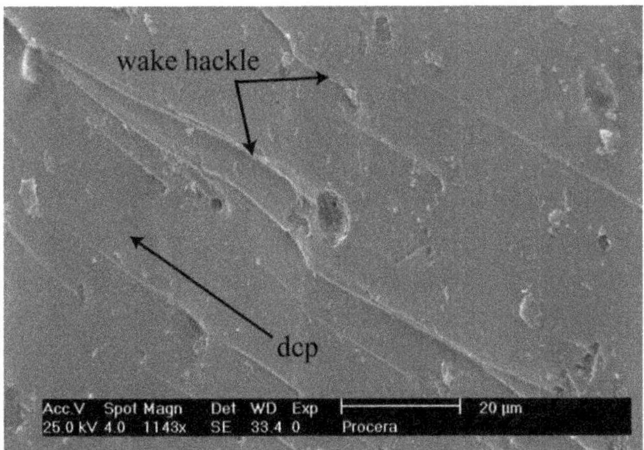

Figure 8.16 SEM image of a Procera all-ceramic crown fractured during service. Examples of wake hackle markings are shown by the joined black arrows. The arrow demarked by dcp shows the direction of crack propagation. This demonstrates how to use wake hackle markings to obtain local crack direction. Source: This figure is courtesy of Scherrer et al. (2008a). Reproduced with permission of Elsevier.

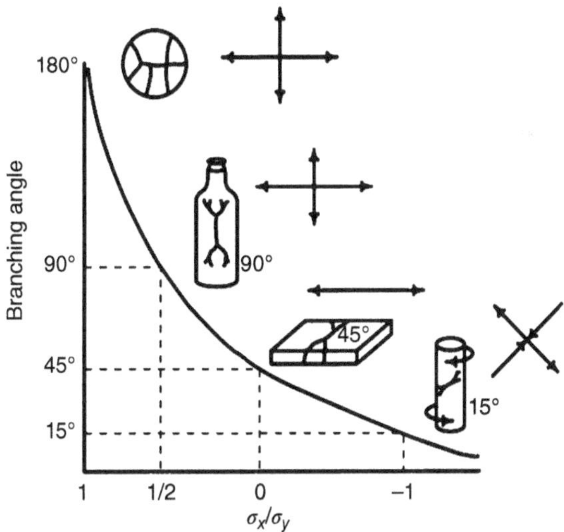

Figure 8.17 Branching angle versus biaxial stress ratios. The macroscopic branching angle is governed by the far-field stress as indicated by σ_x/σ_y. Source: Adapted from Frechette (1990).

secondary crack. Wallner lines can be used to calculate the velocity of a crack front. Notice that the far-field stress governs the branching angle. Thus, the angles of branching provide information about the state and direction of stress application.

When characterizing fracture origins, photographs of the overall sample, of the fracture mirror and surrounding regions including the branching locations (Figure 8.17), if applicable, and of an enlargement of the fracture origin region should be made. The general fractographic procedure is outlined in ASTM Standard C1322 or the Guide to Fractography of Glasses and Ceramics published by NIST (https://www.nist.gov/publications/nist-recommended-practice-guide-fractography-ceramics-and-glasses-0).

Since there is a distribution of crack sizes on any component, there is also a corresponding statistical distribution of strengths. This statistical distribution is usually shown using a probability of failure-strength graph. Fractography can be used to clarify any unusual strength distributions, such as a bimodal strength distribution. For example, it may be assumed that a bimodal strength distribution is due to two flaw types present in the material samples. This assumption may not be correct. Observations of the individual specimens can be used to verify or refute the original suppositions of the relationship between the strength distribution and the flaw types.

A fractographic montage is extremely useful for this purpose (Figure 8.18). The fractographic montage consists of sketches or photographs of the fracture origins with arrows pointing to the location on the probability of failure-strength curve. This type of analysis is especially useful in strength testing and material development.

Fracture surface markings in mixed-mode loading appear different from those in mode I loading. Typical fracture surfaces of a soda-lime-silica glass and a mica glass ceramic in mixed-mode (I/II) loading are shown in Figure 8.19, respectively. Notice that there is an absence of the mist region in mixed-mode loading. Looking very closely at the mixed-mode fracture surfaces (Figure 8.20), we can see that the hackle markings are different and appear as lances. The lancelike features observed on the mixed-mode fracture surfaces of soda-lime-silica glass (Smekal 1953; Sommer 1969; Gopalakrishnan and Mecholsky 2014) are thought to be generated due to the influence

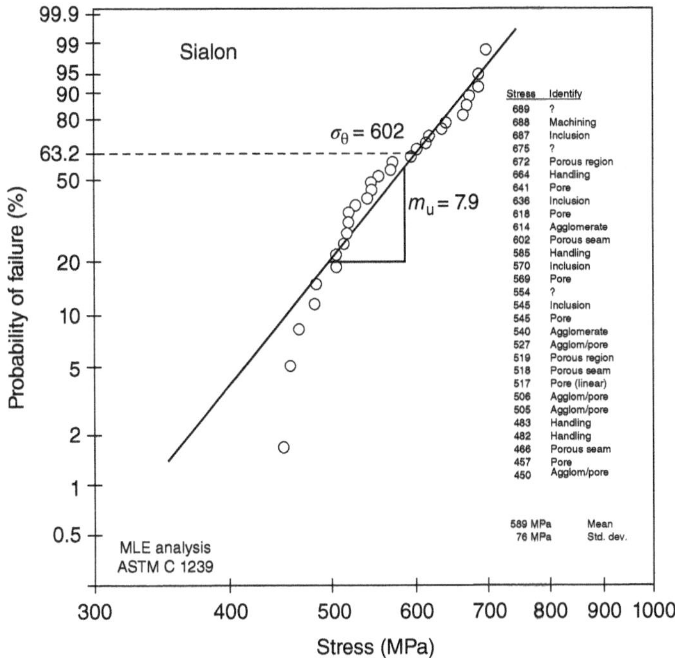

Figure 8.18 Probability of failure as a function of applied stress at failure for sialon. Fractography is used to identify each failure origin and location on the Weibull graph. Source: From Quinn (2007).

of mode I and mode III stress intensities acting at and along the crack front (Pons and Karma 2010). The reason for the lance/hackle markings can be explained by the presence of a mode III component at the tip of the crack. Pons and Karma (2010) show that a "flat parent crack segments into an array of daughter cracks that rotate toward a direction of maximum tensile stress" in a local mixed-mode (I–III) loading condition. They refer to these daughter cracks as "lances," clearly the same as the observed twist hackle markings. Their calculations imply that the creation of the twist hackle features depends on both the applied shear stress and the energy release rate. Other work (Gao and Rice 1986) has also shown that far-field mixed-mode (K_I–K_{II}) loading can result in a mode III shear component at the tip of the crack. Higher order stress intensity terms at the crack tip become significant for mixed-mode conditions.

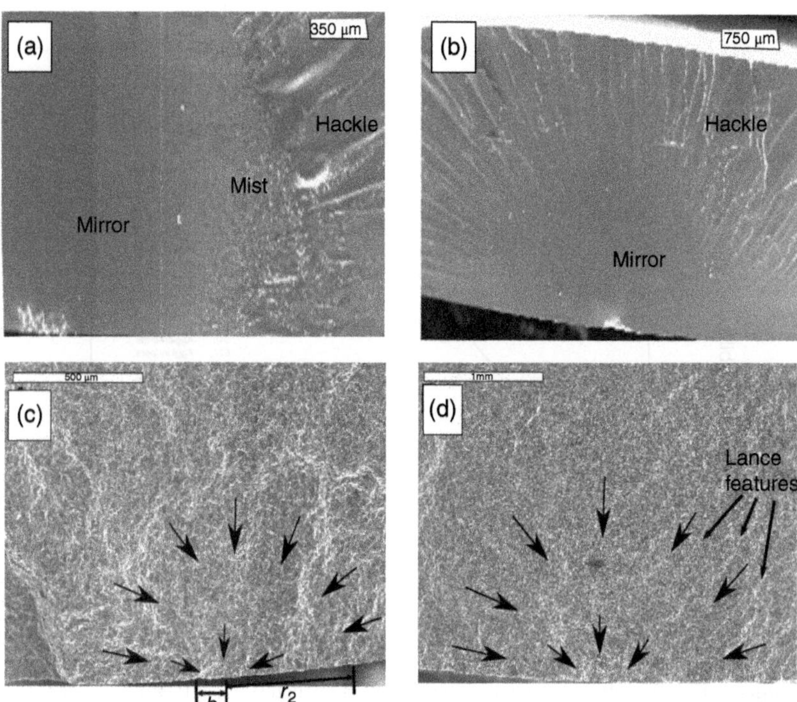

Figure 8.19 Representative fracture surfaces. Soda-lime-silica glass in (a) mode I failure and (b) mixed-mode (I/II) failure. Mica glass ceramic in (c) pure mode I loading and (d) mixed-mode (I/II) loading. For (c) and (d), the inner arrows indicate the critical crack size and the outer arrows indicate the mirror size.

The absence of the mist region and the presence of the characteristic twist hackle markings (or lances) on the fracture surfaces can be used to identify surfaces that fail in combined mode loading. This can be used as a powerful characterization tool to identify chipping failures and sample misorientations while testing. Identification of sample misorientations would help to take into account the orientation effects and make the analysis more accurate.

The twist hackle markings (lances) are observed on the fracture surfaces of single crystals. Tsai and Mecholsky (1992) fractured single crystal silicon flexure bars for two orientations – (i) the {100} tensile surface and the {110} fracture plane and (ii) the {110} tensile surface and {110} fracture plane – and observed the lances and only "mirror"

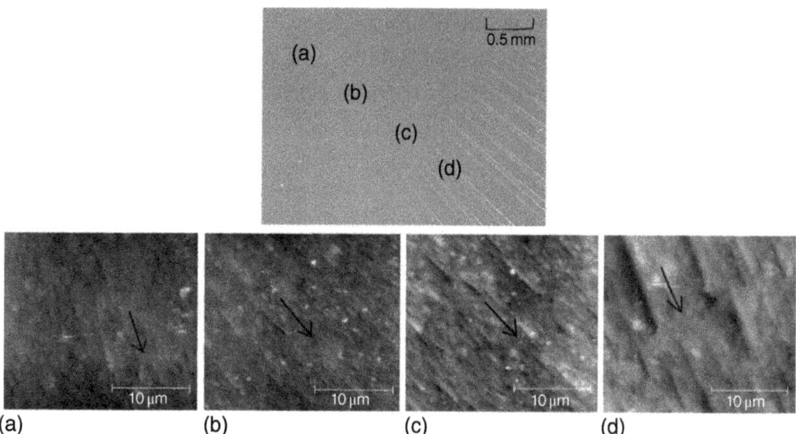

Figure 8.20 AFM images at different regions (marked by letters) on the fracture surface of a soda-lime-silica glass failed in mixed (I/II) mode. (a) Mirror region. (b) Mirror region. (c) Near twist hackle region. (d) Twist hackle region showing lances. The white specks are debris. The arrows in (a)–(d) indicate the crack propagation direction.

and twist hackle features on the fracture surfaces. These lancelike features are seen on fracture surfaces of other single crystals such as sapphire and spinel in various orientations (Quinn 2007). The absence of the mist region and the presence of the lances indicate that the single crystals were subjected to mixed-mode loading in these orientations. Crystal anisotropy results in a natural mode mixity. Thus, fractography provides an important tool to characterize the loading conditions.

The stress intensity at microcrack branching was determined to be a constant for soda-lime-silica glass (Gopalakrishnan and Mecholsky 2014). It is interesting to note that the value of the stress intensity at branching in mixed-mode loading (corresponding to mirror–hackle transition) is similar to that in pure mode I (corresponding to the mirror–mist transition). The stress intensity criterion thus remains valid to explain the crack branching phenomenon in mixed-mode loading conditions. The mirror radius (distance from flaw origin to the mirror–hackle boundary) in mixed-mode loading is related to the far-field stress through the same relation as seen for pure mode I loading (Eq. 8.4). This implies that fractographic principles remain the same in both pure mode I and mixed-mode loading. The fracture stress in mixed mode can be estimated from the knowledge of the stress intensity at branching and the branching radius. Thus, the result of this work supports the evidence that forensic analysis can be used without a priori knowledge of the loading conditions for materials failed in a brittle manner.

FRACTAL GEOMETRY

Fractal geometry is being used in many fields of materials science, physics, chemistry, and engineering because it can be applied to describe shapes and processes that are nonlinear and seemingly complex. Fractal geometry is a non-Euclidean geometry that exhibits self-similarity (or self-affinity) and scale invariance. Self-similarity and scale invariance are characteristics of (scaling) fractals. A self-similar surface is one in which the length scaling is isotropic, i.e. one feature on the surface has a similar shape to another feature on the surface in any direction. In Figure 8.19, the two circled regions in Figure 8.21a are self-similar, i.e. different

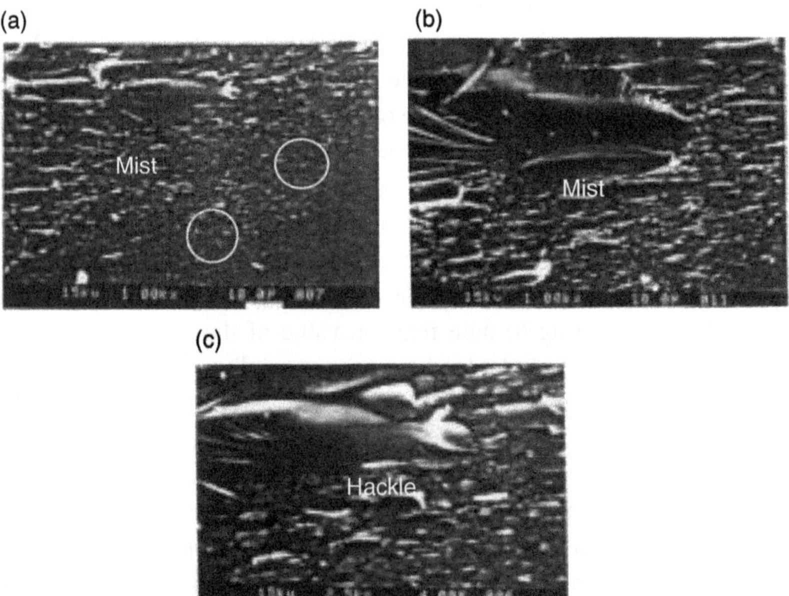

Figure 8.21 Scanning electron micrographs of the mist and hackle regions of a glass fracture surface. The region in (b) is similar to the region in (c). However (c) is the same as (a), only magnified.

regions appear statistically the same. The mathematical statement is that a self-similar object remains invariant under the transformation $(x, y, z) \rightarrow (ax, ay, az)$ where a is a scalar constant. Typically, self-similar structures are also scale invariant, i.e. the features are not only similar in different regions but also similar at different magnifications. In Figure 8.21b the hackle region of a glass fracture surface is shown. In Figure 8.21c, the magnified mist region of Figure 8.19a is shown. Notice that the surfaces in Figure 8.21b and c are almost identical even though they are from different regions and are different magnifications. That feature is a characteristic of scale invariance. A self-affine object has different magnifications in different directions, such as a mountain that has a different scaling relationship on a contour plane versus the elevation scaling. A self-affine object is one that remains invariant under the transformation $(x, y, z) \rightarrow (ax, ay, az^{\zeta})$ where ζ is called a roughness exponent. Fracture is one of the phenomena that have been modeled using fractal geometry.

Fractal objects are characterized by their fractal dimension, D, which is the dimension in which the proper measurement of a fractal object is made. For example, if we had a plane square, then the only dimension in which to make a meaningful measurement is 2, i.e. the area of the square. This same concept is generalized for fractal objects in that they can have non-integer dimensions. Thus, a plane square with "bumps" out of the plane would have dimension, $2.D^*$, where D^* is the fractional part of the fractal dimension and represents the amount of tortuosity out of the plane. An object with a fractal dimension of 2.1 ($D^* = 0.1$) would be relatively flat, and an object with $D = 2.9$ ($D^* = 0.9$) would be almost a volume-filling object. If the same relatively flat "bumpy" plane (with $D = 2.1$) was measured by a contour line, then the fractal dimension would be $D = 1.1$, but D^* would still be 0.1. If we examine the fracture surface of a brittle material, for example, an inorganic glass, at several length scales, we find that the "smooth mirror" region is not smooth. Figure 8.22 shows an atomic force microscope image of the mirror region of a soda-lime-silica glass showing surface structure even in the mirror region. Thus, the mirror is not smooth at the atomic scale!

There are several techniques available to measure the fractal dimension. We recommend the slit-island analysis (SIA) technique, first developed by Mandelbrot et al. (1984). The technique was slightly

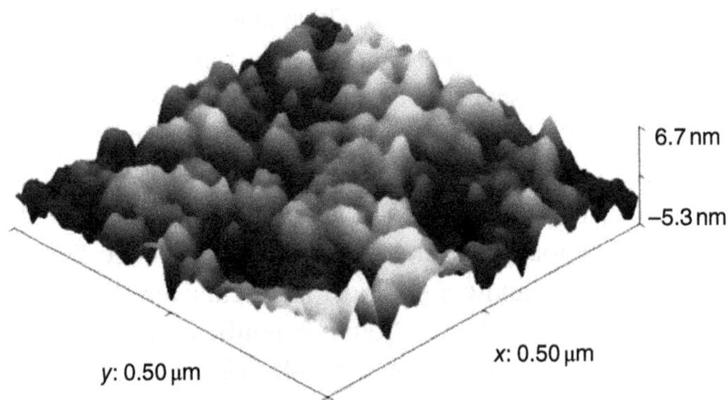

Figure 8.22 Atomic force microscope image of the mirror region of soda-lime-silica glass. Notice that it is not smooth as usually appears at greater magnification. Source: Smith et al. (2009). Reproduced with permission of Springer Nature.

modified (Hill et al. 2001) to make measurements on replicas of the fracture surface. Thus, several images of the fracture surface can be used, and the fracture surface is preserved. The procedure is as follows: (i) The fracture surface of the specimen is thoroughly cleaned with ethanol and rinsed with distilled water. (ii) A negative impression is made of the fracture surface using a polyvinylsiloxane impression material (Kerr Manufacturing Co., Romulus, MI). During this process, a thin layer of the material is slowly applied to minimize the trapped air between the fracture surface and the impression. If bubbles or defects are present, the impression should be discarded, and a new one created. More impression material is then added to fill in behind the initial thin layer. The impression is then allowed to polymerize completely before it is carefully separated from the fracture surface. (iii) A positive epoxy replica is created from the impression according to the manufacturer's instructions (Leco epoxy kit 811-161, St. Joseph, MI). This impression should be coated for about four minutes using gold/palladium to produce a thick high contrast layer. A subsequent layer of epoxy is then poured over the coated replica creating a sandwich-type specimen (Figure 8.23). Several replicas can be produced from each impression (Scherrer et al. 2008b). For the modified SIA technique, the specimen replica, as shown in Figure 8.23, is polished as close to parallel to the fracture surface as possible until "islands" appeared. The surfaces are then polished through 1 μm alumina slurry, preferably by hand, to

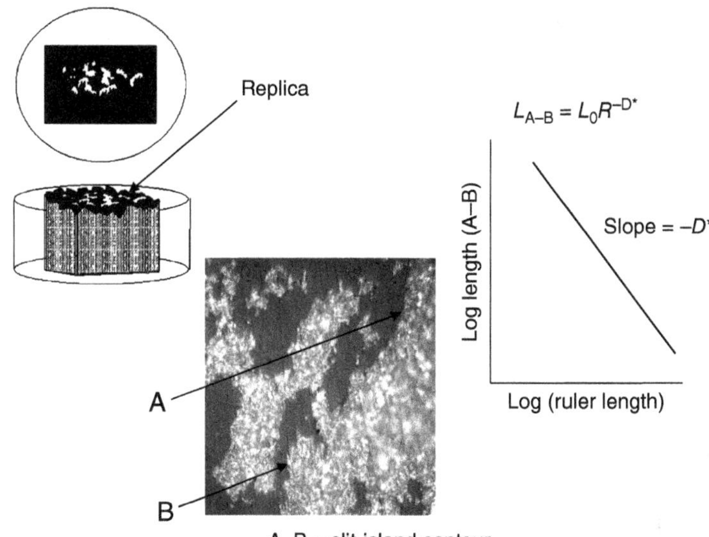

Figure 8.23 Schematic summarizing slit-island analysis (SIA) technique. The Richardson equation ($L_{A-B} = L_0R^{-D^*}$ where R is the ruler length) is used to determine the fractal dimensional increment, D^*. L_0 is a constant.

produce a sharper boundary. A series of 8–10 photographs should be taken at a magnification of 400× of part of the coastline of one of the islands. These photographs are arranged in a collage where the coastline is measured using dividers in order to apply the Richardson equation shown in Figure 8.23. The SIA technique is modified in that the original presentation (Mandelbrot et al. 1984) measured the area and perimeter of a selected island at different magnifications. We are suggesting measuring part of the perimeter (coastline) using several different ruler lengths.

Fundamental relationships can be derived between the fractal dimensional increment, D^*, and the fracture toughness of a material in the form of the critical stress intensity factor, K_{IC}:

$$K_{IC} = Ea_0^{1/2}D^{*1/2} = Y(\theta)\sigma_f c^{1/2} \tag{8.10}$$

where E is Young's modulus and a_0 is a material parameter having the units of length (Mecholsky 1996; West et al. 1999). The relationship between K_{IC} and D^* is based on experimental observations (Figure 8.24), and the relationship between K_{IC} and c is based on fracture mechanics

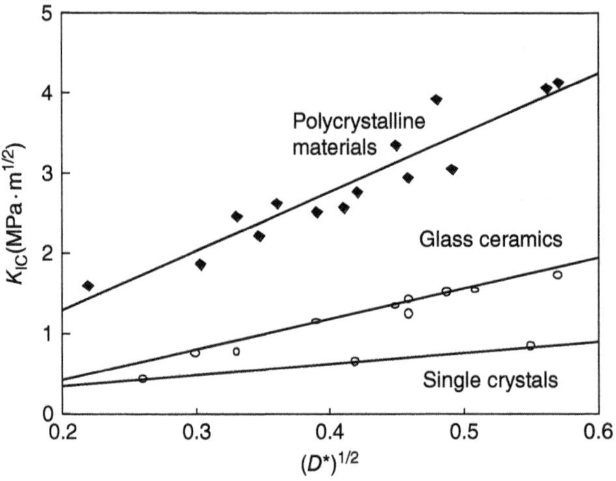

Figure 8.24 Fracture toughness versus the fractal dimensional increment, D^*. The results group into classes of materials that behave in a similar manner such as single crystals, glass ceramics, and fine-grain polycrystalline materials with few toughening mechanisms.

and experimental confirmation. The measurement of D^* is an average property of the entire fracture surface and a measure of its tortuosity.

It can be shown that (Mecholsky and Freiman 1991)

$$\frac{r_1}{c} = \frac{1}{D^*} \tag{8.11}$$

The agreement of experimental data with Eq. (8.11) shown in Figure 8.25 implies that there is scaling between the crack size and the point of branching. Furthermore, the combination of Eqs. (8.3) and (8.4) shows that the ratio of the energy of crack initiation to that of crack branching is related to r_j/c; there is also scaling between the energy of crack initiation and of crack branching. The relationship in Eq. (8.11) will still be valid for the other values of j, but will be numerically different due to a different proportionality constant.

Thus, analyses considering an energy scaling or a structural (geometric) scaling are equivalent. Moreover, the scaling is linear in energy or geometry. Passoja (1988) has suggested that this scaling is a general phenomenon that is observed in many materials and is related to the atomic structure.

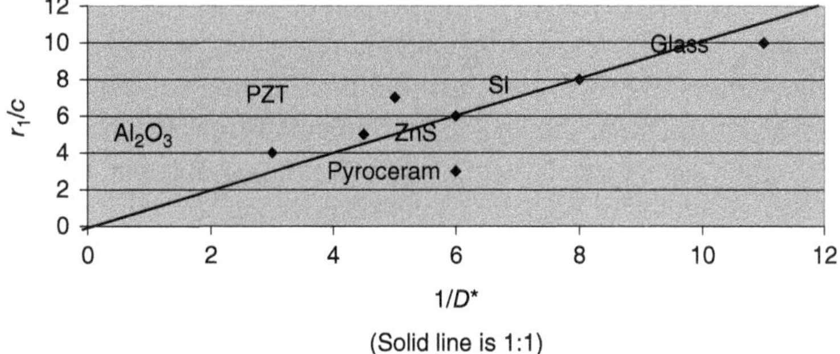

Figure 8.25 Mirror-to-crack size ratio as a function of $1/D^*$. Silicon (Si) is single crystalline. The rest of the materials are polycrystalline except for glass that is a silicate-based glass. This graph shows the relationship between fractography and fractal geometry of the fracture surface. Source: Mecholsky and Freiman (1991). Reproduced with permission of John Wiley & Sons.

Fractal dimensions for fracture surfaces of glasses, ceramics, glass ceramics, single crystals, intermetallics, and polymers have been measured. Atomic force microscopy studies have shown that the fracture surface is not smooth at the atomic level for many materials and has the same tortuosity at the atomic level as at the macroscopic level (Guin and Wiederhorn 2004; Smith et al. 2009). Thus, the crack front is neither smooth nor continuous. This finding that the "mirror" region is not smooth affects our judgment in describing fracture surfaces as related to fracture testing results. It should remind us to use many tools to observe fracture surfaces and that fracture surfaces should be observed at many length scales for accurate interpretation. Fractography can aid us in understanding the results of testing materials for a measurement of their mechanical properties. When we analyze unexpected results, we should always apply as many available techniques as possible. The study of fractal geometry suggests we need to examine even finer length scales to understand the fundamental behavior of brittle materials.

SUMMARY

Fracture in brittle materials mainly occurs by the initiation of a defect in a critical location of greatest stress. Usually, this defect is a semielliptical crack or can be modeled as a semielliptical crack. Linear elastic fracture mechanics (LEFM) equations can be used to characterize the stress and crack size needed for initiation of crack propagation. After the crack starts to propagate, there are three regions of crack growth known as mirror, mist, and hackle. Each region is rougher than the previous region. The crack starts in a relatively smooth manner and jumps to a rougher region at boundaries between the mirror–mist and mist–hackle and finally at the point between the end of the hackle region and macroscopic crack branching. There are equations that govern the transition between these regions analogous to the fracture mechanics (LEFM) equations. Thus, there is a constant relationship between the crack size and mirror radii for each material. Quantitative fractography utilizes the features on the fracture surfaces by measuring the transitions between the start of crack propagation and the transitions of these characteristic regions. Useful equations can be developed that can identify whether or not local residual stress existed at the start of propagation, whether there was slow crack growth due to environmental or other effects, and the existence of any global residual stress in the material at the time of fracture. There is much information to be obtained in testing ceramic material by observing the fracture surface. The information is free, i.e. it exists as a consequence of the fracture process. The use of this free information is to be determined by the observer.

QUESTIONS

1. You tested a rectangular bar (10 mm × 20 mm × 100 mm) of NC132 silicon nitride in four (1/4)-point flexure, and it failed at a load of 200 kN. You examine the fracture surface and measure the critical crack size on the tensile surface. The crack measured 30 μm in depth and 100 μm in width. The mirror radii were symmetrical, and the measurements along the tensile surface were r_1 = 150 μm, r_2 = 200 μm, and r_3 = 250 μm. From your measurements, what can you say about the failure? [Hint: Is there any residual stress? Did it fail rapidly?]
2. Glass fibers are often tested on mandrels. The fibers are wrapped around the mandrel, and then two mandrels are rotated at slightly different speed to produce tension between the mandrel. A silica glass fiber was fractured while being stressed between a mandrel, and the stress was calculated from the failure load and diameter to be 300 MPa. You examined the fracture surface using a scanning electron microscope and determined the crack size to

be 9 μm. The mirror–mist radius was 85 μm. Is the test valid? Is the stress an accurate representation? Why or why not?

3. Sketch the expected fracture surface on a rectangular bar fractured in four-point flexure from a critical semicircular crack of 50 μm radius in the center of the bar for the following conditions: (i) fast fractured in an annealed and (internal) stress-free state, (ii) fast fractured with a residual compressive stress on the surface that is 10 μm deep, (iii) slowly loaded so that there was slow crack growth due to the environment. Use the same length scale for each of the sketches.
4. Calculate the critical stress intensity factor for a semielliptical crack that is 50 by 150 μm and was subjected to a tensile stress of 100 MPa. (i) Use Eq. (8.1). (ii) Use Eq. (8.3). Are these values significantly different? Why or why not?
5. Alumina specimens contain flaws introduced during processing; these flaws are approximately the grain size, d. Graph the fracture stress vs. grain size for grains below 200 μm. Assume the fracture toughness is $4.5\,\text{MPa}\cdot\text{m}^{1/2}$. Also, assume $d = 2c$.
6. Show that $(ab)^{1/2} = c$, where a is the semiminor axis, b is the semimajor axis of an ellipse, and c is the radius of a semicircle. What is the condition for this to be true?
7. Examining the fracture surface of a soda-lime-silica glass plate that failed. How would you locate the origin? How would you determine the stress at which it failed? Could you determine the state of stress at failure? If so, how? If not, why not?

REFERENCES

Anderson, T.L. (2005). *Fracture Mechanics: Fundamentals and Principles*, 3e. Boca Raton, FL: CRC Press.

Bar-on, I. (2001). Quantitative analysis. Fractography of brittle materials. In: *Encyclopedia of Materials: Science & Technology* (ed. K.H. Buschow, R. Cahn, M. Flemings, et al.), 3260. Elsevier.

Beauchamp, E.K. (1996). Mechanisms for hackle formation and crack branching. In: *Fractography of Glasses and Ceramics III*, Ceramic Transactions, vol. 64 (ed. J.R. Varner, V.D. Frechette and G.D. Quinn), 409–445. Westerville, OH: The American Ceramic Society.

Beauchamp, E.K. and Matalucci, R.V. (1998). *Dynamics of Window Glass Fracture in Explosions*, SAND98-0598. Albuquerque, NM: Sandia National Laboratories.

Choi, S.R. and Gyekenyesi, J.P. (1998). *Crack Branching and Fracture Mirror Data of Glasses and Advanced Ceramics*, NASA/TM – 1998-206536. Linthicum Heights, MD: NASA Center for AeroSpace Information.

Conway, J.C. Jr. and Mecholsky, J.J. Jr. (1989). Use of crack branching data for measuring near-surface residual stresses in tempered glass. *J. Am. Ceram. Soc.* 72 (9): 1584–1587.

Danzer, R., Lube, T., Supancic, P., and Damani, R. (2008). Fracture of ceramics. *Adv. Eng. Mater.* 10 (4): 275–298.

De Freminville, C. (1914). Recherches sur la fragilité l'éclatement. *Rev. Met.* 11: 971–1056.

Frechette, V.D. (1990). *Failure Analysis of Brittle Materials*, Advances in Ceramics, vol. 28, 135. Westerville, OH: The American Ceramic Society.

Freiman, S.W., Mecholsky, J.J., Rice, R.W., and Wurst, J.C. (1975). Influence of microstructure on crack propagation in ZnSe. *J. Am. Ceram. Soc.* 58 (9–10): 406.

Freiman, S.W., Mecholsky, J.J., and Becher, P.F. (1991). Fractography: a quantitative measure of the fracture process. In: *Fractography of Glasses and Ceramics II*, Ceramic Transactions, vol. 17 (ed. V.D. Frechette and J. Varner), 55–78. Westerville, OH: The American Ceramic Society.

Gao, H. and Rice, J.R. (1986). Shear stress intensity factors for a planar crack with slightly curved front. *J. Appl. Mech.* 53: 774–778.

Gopalakrishnan, K. and Mecholsky, J.J. Jr. (2014). Quantitative fractography of mixed-mode fracture in glass and ceramics. *J. Eur. Ceram. Soc.* 34 (14): 3247–3254.

Guin, J.P. and Wiederhorn, S.M. (2004). Fracture of silicate glasses: ductile or brittle? *Phys. Rev. Lett.* 92: 215502–215506.

Hill, T.J., Della-Bona, A., and Mecholsky, J.J. Jr. (2001). Establishing a protocol for measurements of fractal dimensions in brittle materials. *J. Mater. Sci.* 36: 2651–2657.

Kirchner, H.P. (1986). Brittleness dependence of crack branching in ceramics. *J. Am. Ceram. Soc.* 69 (4): 339–342.

Kirchner, H.P. and Conway, J.C. Jr. (1988). Fracture mechanics of crack branching in ceramics. In: *Fractography of Glass and Ceramics*, Advances in Ceramics, vol. 22 (ed. J.R. Varner and V.D. Frechette), 187–213. Westerville, OH: The American Ceramics Society.

Kirchner, H.P. and Kirchner, J.W. (1979). Fracture mechanics of fracture mirrors. *J. Am. Ceram. Soc.* 62 (3–4): 198–202.

Kulawansa, D.M., Jensen, L.C., Langford, S.C., and Dickinson, J.T. (1994). STM observations of the mirror region of silicate glass fracture surfaces. *J. Mater. Res.* 9 (2): 476–485.

Mandelbrot, B.B., Passoja, D.E., and Paulley, A.J. (1984). Fractal character of fracture surfaces of metals. *Nature* 308: 721.

Marshall, D.B. and Lawn, B.R. (1986). Indentation of brittle materials. In: *Microindentation Techniques in Materials Science & Engineering*, ASTM STP

889 (ed. P.J. Blau and B.R. Lawn), 26–46. Philadelphia, PA: American Society for Testing and Materials.

Marshall, D.B., Lawn, B.R., and Mecholsky, J.J. Jr. (1980). Effect of residual contact stresses on mirror/flaw size relations. *J. Am. Ceram. Soc.* 63: 7–8.

Maurer, R.D., Miller, R.A., Smith, D.D., and Trondsen, J.C. (1974). Optimization of Optical Wave Guides-Strength Studies. Corning Glass Works Technical Report, ONR Contract N00014-73-C-0293.

Mecholsky, J.J. Jr. (1991). Quantitative fractography: an assessment. In: *Fractography of Glasses and Ceramics II*, Ceramic Transactions, vol. 17 (ed. V.D. Frechette and J.P. Varner), 413–451. Westerville, OH: The American Ceramic Society.

Mecholsky, J.J. Jr. (1992). Section 9: Failure analysis. Fractography of optical fibers. In: *ASM Engineered Materials Handbook, 4, Ceramics and Glasses* (ed. S. Schneider). Materials Park, OH: The American Society for Metals.

Mecholsky, J.J. Jr. (1996). Fractography, fracture mechanics and fractal geometry: an integration. In: *Fractography of Glasses and Ceramics III*, Ceramic Transactions, vol. 64 (ed. J.P. Varner, V.D. Frechette and G.D. Quinn), 385–393. Westerville, OH: The American Ceramic Society.

Mecholsky, J.J. Jr. (2001). Fractography: brittle materials. In: *Encyclopedia of Materials: Science and Technology* (ed. K.H.J. Buschow, R.W. Cahn, M.C. Flemings, et al.), 3257–3265. Oxford, UK: Pergamon.

Mecholsky, J.J. Jr. and Freiman, S.W. (1991). Relationship between fractal geometry and fractography. *J. Am. Ceram. Soc.* 74 (12): 3136–3138.

Mecholsky, J.J. Jr., Rice, R.W., and Freiman, S.W. (1974). Prediction of fracture energy and flaw size in glasses from mirror size measurements. *J. Am. Ceram. Soc.* 57: 10.

Mecholsky, J.J. Jr., Rice, R.W., and Freiman, S.W. (1976). Fracture surface analysis of ceramics. *J. Mater. Sci.* 11: 1310–1319.

Mecholsky, J.J. Jr., Freiman, S.W., and Morey, S.M. (1977a). Fractographic analysis of optical fibers. *Bull. Am. Ceram. Soc.* 56 (11): 1016–1017.

Mecholsky, J.J. Jr., Freiman, S.W., and Rice, R.W. (1977b). Effect of grinding on flaw geometry and fracture of glass. *J. Am. Ceram. Soc.* 60 (3–4): 114–117.

Mecholsky, J.J. Jr., Gonzales, A.C., and Freiman, S.W. (1979). Fractographic analysis of delayed failure in soda lime glass. *J. Am. Ceram. Soc.* 62 (11–12): 577–580.

Michalske, T.A. (1979). Fractography of slow fracture. In: *Fractography of Ceramic and Metal Failures*, ASTM STP 827 (ed. J.J. Mecholsky and S.R. Powell), 136–150. Philadelphia, PA: American Society for Testing and Materials.

Morrell, R. (2006). Fracture toughness testing for advanced technical ceramics: internationally agreed good practice. *Adv. Appl. Ceram.* 105 (2): 88–98.

Passoja, D.E. (1988). Fundamental relationships between energy and geometry in fracture. In: *Fractography of Glasses and Ceramics* (ed. V.D. Frechette and J.P. Varner), 101–126. Westerville, OH: The American Ceramic Society.

Pons, J.A. and Karma, A. (2010). Helical crack-front instability in mixed-mode fracture. *Nature* 464: 85–89.

Preston, F.W. (1926). A study of the rupture of glass. *J. Soc. Glass Technol.* 10 (39): 234–269.

Quinn, J.B. (1999). Extrapolation of fracture mirror and crack-branch sizes to large dimensions in biaxial strength tests of glass. *J. Am. Ceram. Soc.* 82 (8): 2126–2132.

Quinn, G.D. (2007). *Fractography of Ceramics and Glasses.* A NIST Recommended Practice Guide, Special Publication 960-16. Washington, DC: National Institute of Standards and Technology.

Quinn, G.D. (2016). *Fractography of Ceramics and Glasses.* A NIST Recommended Practice Guide, Special Publication 960-16e2. Washington, DC: National Institute of Standards and Technology.

Quinn, G.D. and Swab, J.J. (1996). Fractography and estimates of fracture origin size from fracture mechanics. *Ceram. Eng. Sci. Proc.* 17 (3): 1–58.

Quinn, G.D., Swab, J.J., and Motyka, M.J. (2003). Fracture toughness of a toughened silicon nitride by ASTM C 1421. *J. Am. Ceram. Soc.* 86 (6): 1043–1045.

Quispe-Cancapa, J.J., Lopez-Cepero, J.-M., de Arellano-Lopez, A.R., and Martınez-Fernandez, J. (2008). High-temperature fracture toughness of chromium-doped sapphire fibers. *J. Am. Ceram. Soc.* 91 (12): 3994–4002.

Raju, I.S. and Newman, J.C. (1979). Stress intensity factors for a wide range of semielliptical surface cracks in finite thickness plates. *Eng. Fract. Mech.* 11: 817–829.

Randall, P.N. (1967). *Plane Strain Crack Toughness Testing of High-Strength Metallic Materials*, ASTM STP 410 (ed. W.F. Brown), 88–126. Philadelphia, PA: American Society for Testing and Materials.

Rice, R.W. (1984). Ceramic fracture features, observations, mechanisms and uses. In: *Fractography of Ceramic and Metal Failures*, ASTM STP 827 (ed. J.J. Mecholsky Jr. and S.R. Powell Jr.), 5–102. Philadelphia, PA: American Society for Testing and Materials.

Rice, R.W. (1994). Porosity effects on machining direction – strength anisotropy and failure mechanisms. *J. Am. Ceram. Soc.* 77: 2232–2236.

Rice, R.W., Freiman, S.W., and Mecholsky, J.J. Jr. (1980). Effect of flaw to grain size ration on fracture energy of ceramics. *J. Am. Ceram. Soc.* 63: 129–136.

Rossmanith, H.P. (1980). Crack Branching in Brittle Materials: Part I. Analytical Aspects. University of Maryland Research Report, Photomechanics Lab, University of Maryland, College Park, MD.

Salem, J.A. and Jenkins, M.G. (2002). Estimating bounds on fracture stresses determined from mirror size measurements. *J. Am. Ceram. Soc.* 85 (3): 706–708.

Sanford, R.J. (2003). *Principles of Fracture Mechanics.* Upper Saddle River, NJ: Prentice Hall.

Scherrer, S.S., Quinn, G.D., and Quinn, J.B. (2008a). Fractographic failure analysis of a Procera® AllCeram crown using stereo and scanning electron microscopy. *Dent. Mater.* 24 (8): 1107–1113.

Scherrer, S.S., Quinn, J.B., Quinn, G.D., and Anselm Wiskott, H.W. (2008b). Fractographic ceramic failure analysis using the replica technique. *Dent. Mater.* 23 (11): 1397–1404.
Shand, E.B. (1959). Breaking strength of glass determined from dimensions of fracture mirrors. *J. Am. Ceram. Soc.* 42 (10): 474–477.
Smekal, A. (1953). Zum bruchvorgang bei sprödem stoffverhalten unter ein-und mehrachsigen beanspruchungen. *Oesterr Ing Arch* 7: 49–70.
Smith, R.L. III, Mecholsky, J.J. Jr., and Freiman, S.W. (2009). Estimation of fracture energy from the work of fracture and fracture surface area: I. Stable crack growth. *Int. J. Fract.* 156: 97–102.
Sommer, E. (1969). Formation of fracture "lances" in glass. *Eng. Fract. Mech.* 1: 539–546.
Taskonak, B., Yan, J., Mecholsky, J.J. Jr. et al. (2008). Fractographic analyses of zirconia based fixed partial dentures. *Dent. Mater.* 24 (8): 1077–1082.
Tsai, Y.L. and Mecholsky, J.J. Jr. (1991). Fractal fracture in single crystal silicon. *J. Mater. Res.* 6 (6): 1248–1263.
Tsai, Y.L. and Mecholsky, J.J. Jr. (1992). Fracture mechanics description of fracture mirror formation in single crystals. *Int. J. Fract.* 57: 167–182.
Wallner, H. (1939). Linienstrukturen an Bruchflachen. *Z. Phys.* 114: 368–378.
West, J.K., Mecholsky, J.J. Jr., and Hench, L.L. (1999). The quantum and fractal geometry of brittle fracture. *J. Noncryst. Solids* 260: 99–108.

CHAPTER 9

Microstructural Effects

INTRODUCTION

Because of the relatively low fracture toughness of ceramics, flaws are frequently of the same size as the grains in the material. Therefore, a significant factor in the interpretation of fracture data in brittle materials is their microstructure, i.e. grain size, shape, and orientation. The microstructure can affect fracture in many ways including increasing the material's crack growth resistance.

GRAIN SIZE AND ORIENTATION EFFECTS

Depending on the depth of the surface defects and the possible existence of large grains, a flaw can be embedded in a single crystal or encompass many grains, raising the issue of whether fracture toughness values obtained with large cracks are a valid predictor of the behavior of small cracks. Artificial cracks, which can be many times the size of the grains, can yield resistances to crack growth many times that of small flaws. Cracks, which encompass many grains, will be more resistant to growth because of toughening mechanisms that do not apply to smaller cracks.

The Fracture of Brittle Materials: Testing and Analysis, Second Edition.
Stephen W. Freiman and John J. Mecholsky, Jr.

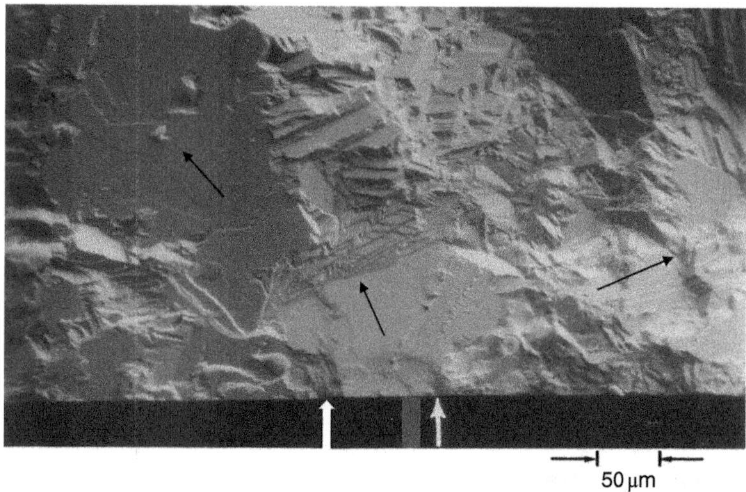

Figure 9.1 Fracture surface of large-grained zinc selenide. The two white arrows indicate the edges of the critical crack boundaries. The black arrows indicate the direction of crack propagation indicated by the fracture steps fanning out from the origin. Source: Freiman et al. (1975). Reproduced with permission of John Wiley & Sons.

An example of the potential effects of flaw size relative to grain size on fracture toughness is shown in Figures 9.1 and 9.2. Figure 9.1 shows the fracture initiation site in a polycrystalline ZnSe. It is clear that fracture initiated from a flaw wholly contained within a large grain of the material. In Figure 9.2 it is seen that a graph of flexural strength versus the size of the measured flaws within large grains (cf. Eq. 2.1) yields a value of fracture energy representing that of a single crystal of ZnSe rather than the polycrystalline value.

The transition from fracture governed by the individual grain to that determined by the polycrystalline microstructure occurs over a range of flaw/grain sizes (Figure 9.3). Figure 9.4 illustrates that the transition from single-crystal governed failure to polycrystalline is material dependent.

An important factor in the transition is the number of cleavage planes and their orientation with respect to one another. Despite the fact that the polycrystalline material surrounding the large grain can be significantly more resistant to crack growth, if the crack achieves a critical size within the grain, this increased resistance will not prevent failure.

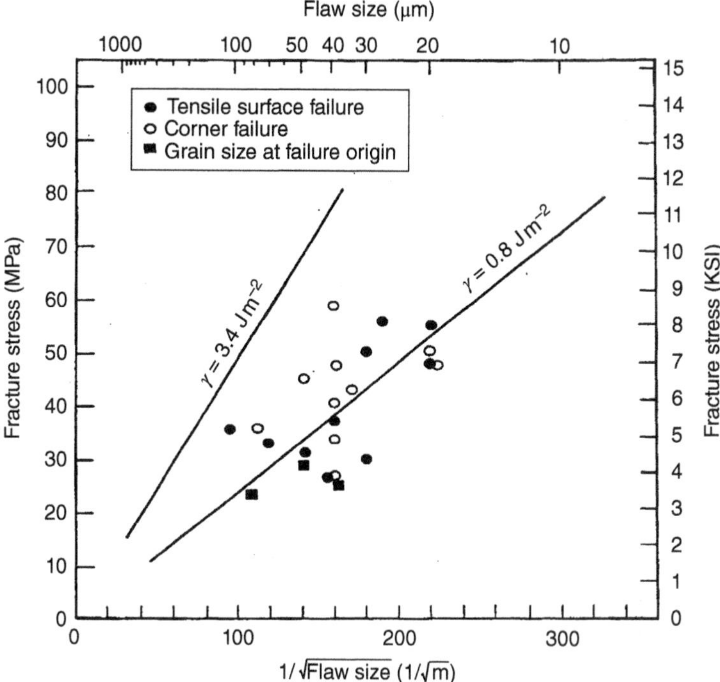

Figure 9.2 Fracture strength of the ZnSe materials as a function of measured flaw sizes. Lines represent values of fracture energy for polycrystalline material ($3.4\,J\,m^{-2}$) and single-crystal ZnSe ($0.8\,J\,m^{-2}$) used in Eq. (8.3). Source: Freiman et al. (1975). Reproduced with permission of John Wiley & Sons.

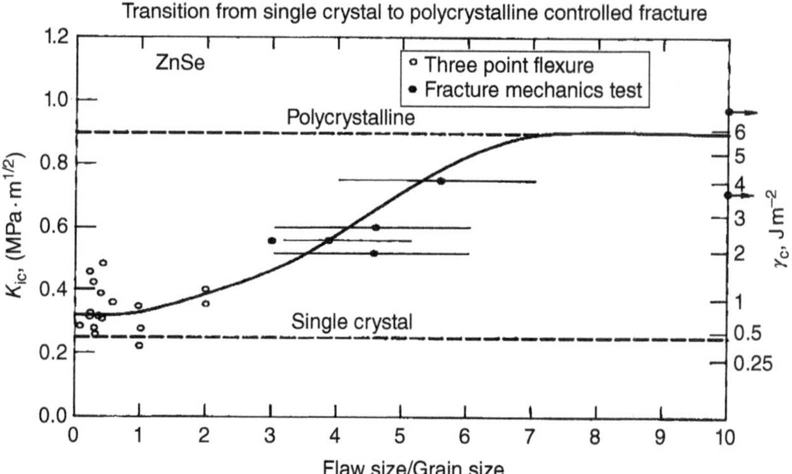

Figure 9.3 Fracture toughness of ZnSe as a function of the flaw size-to-grain size ratio. There is a gradual transition to the polycrystalline value over several grains. Source: Rice et al. (1980). Reproduced with permission of John Wiley & Sons.

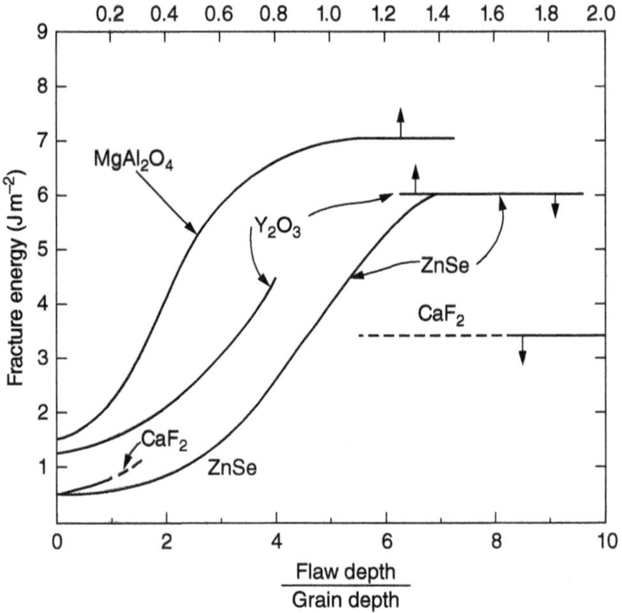

Figure 9.4 Transition in fracture energy for a number of materials as a function of flaw size/grain size. Source: Rice et al. (1981). Reproduced with permission of John Wiley & Sons.

This statement appears to be true regardless of how close to the large-grain boundary the initial flaw exists. Any possible dynamic effects resulting from a rapidly moving crack seem to be minimal. There have been numerous discussions regarding the behavior of crack propagation behavior within large grains (Singh et al. 1979; Rice et al. 1981). Although there is still some disagreement, most investigators agree that if the grain size ratio is 8–10 times the crack size, no arrest will occur at the grain boundary. The disagreement occurs when the crack size is near the boundary of the grain (Singh et al. 1979) or there is substantial slow crack growth within the grains (Kirchner and Ragosta 1980).

From a crack perspective, we can view the transition schematically as shown in Figure 9.5. When the crack is contained within the grain, it propagates along a cleavage plane. Once the growing crack reaches the large grain boundary, in order for it to continue to propagate on this plane in the polycrystalline region, segments of the crack must alter their orientation to conform to that of the grain over that segment.

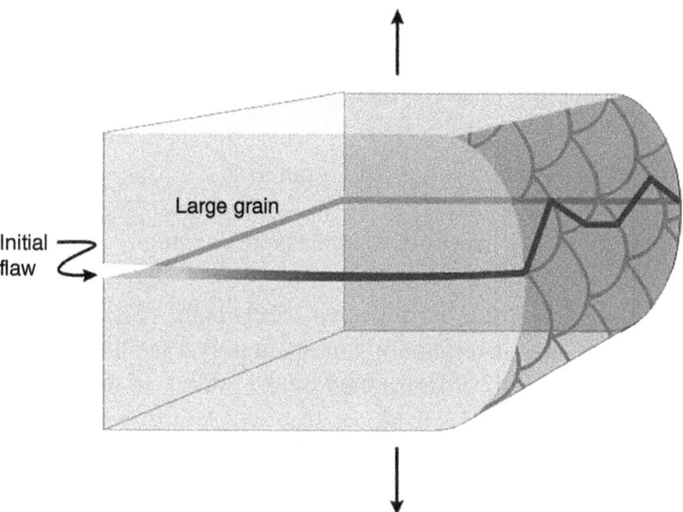

Figure 9.5 Schematic showing the relationship between a flaw growing in a large grain under stress and the segments of the growing crack in the adjoining smaller grains. Source: Freiman (2002).

The stress intensity factor needed to extend such misaligned segments can be calculated from mixed-mode models of crack growth (see Chapter 4). One such expression assumes a coplanar extension of the crack under mixed-mode loading (Paris and Sih 1965). The expression governing such a crack is

$$K_{IC} = K_I \left[1 + \left(\frac{K_{II}}{K_I} \right)^2 \right]^{1/2} \tag{9.1}$$

where K_{IC} is the fracture toughness and K_I and K_{II} are the mode I and mode II stress intensity factors whose values can be calculated from Eq. (9.2a and b):

$$K_I = \Omega_I \sigma c^{1/2} \cos^2 \theta \tag{9.2a}$$

$$K_{II} = \Omega_{II} \sigma c^{1/2} \sin\theta \cos\theta \tag{9.2b}$$

In these expressions, σ is the applied stress; c is the crack radius; $\Omega_I = 2m/\pi^{½} = 1.35$; $\Omega_{II} = 4/\pi^{½}(2-\nu) = 1.29$, where $m = 1.2$ for semicircular cracks; and ν is Poisson's ratio, taken to be 0.25. θ is the angle between the applied stress and the initial crack plane. Combining the above equations and taking $\Omega_I = \Omega_{II}$, then

$$K_{IC} = K_{IC(SC)}\left[1 + \tan^2\theta\right]^{1/2} \tag{9.3}$$

where $K_{IC(SC)}$ is the fracture toughness of a single crystal of the material.

Assuming that there is only one crack path, namely, the single cleavage plane, then in a randomly oriented polycrystalline microstructure, an average θ will be determined by the grain size. For self-similar grain shapes, the area of fracture surface will be independent of grain size. However, finer grain size material will give rise to smaller values of this angle.

While other possible toughening mechanisms such as crack bridging, formation of subsidiary cracks, grain boundary fracture, etc. could come into play, Eq. (9.3) provides the minimum extra energy that would be required over and above that needed to grow a crack through a single grain. Taking plausible values of θ, one can calculate expected values of $K_{IC}/K_{IC(SC)}$ as shown in Table 9.1. Based on the calculated values of $K_{IC}/K_{IC(SC)}$ shown in Table 9.1, it appears that mixed-mode fracture within each grain as expressed in Eq. (9.3) can account for the observed increases in toughness as a crack grows out of an isolated large grain.

Table 9.1
Ratio of $K_{IC}/K_{IC(SC)}$ as a Function of Misorientation Angle, θ, of Cleavage Plane Segments

θ (°)	$K_{IC}/K_{IC(SC)}$
0	1
30	1.2
60	2
70	2.9
80	5.7

TOUGHENING MECHANISMS

In addition to grain orientation, there are other mechanisms based on a materials microstructure that can lead to larger fracture resistance.

Crack Deflection

One of the simplest possible toughening mechanisms is the increase in fracture path due to grain-to-grain misorientation along the crack front. A crack propagating along an easy fracture path in one grain, or along a grain boundary, must change direction in order to grow through or around the next differently oriented grain. The degree of deflection determines the increase in toughness.

The model for this mechanism was derived based on the tilt and twist of a crack (Faber and Evans 1983), which affects the localized stresses. Their model suggests that elongated grains should have a more significant effect on toughness than equiaxed ones and leads to predicted increases in K_{IC} of up to 40%. Experimental data obtained on silicon nitride and a baria-silica glass ceramic generally agreed with this model. Other work on a series of crystallized glasses (Hill et al. 2000) confirms the possibility of such large toughness increases due to crack deflection.

Microcracking

It has been observed that the fracture toughness of polycrystalline ceramics having a cubic crystal structure, e.g. SiC and spinel ($MgAl_2O_4$) (Figure 9.6), is independent of grain size, while that for non-cubic ceramics, e.g. Al_2O_3, (Figure 9.7) passes through a maximum at a particular grain size. Microcracking, caused by internal stress in the material, plays the significant role in this latter behavior.

Stresses can arise in polycrystalline ceramics due to solid-state phase transformations or to anisotropy in the thermal expansion coefficient of the material. The existence of thermal expansion anisotropy in all non-cubic crystals means that cooling these materials from processing temperatures will give rise to differential contraction at different grain facets, leading to localized stresses. While these stresses are

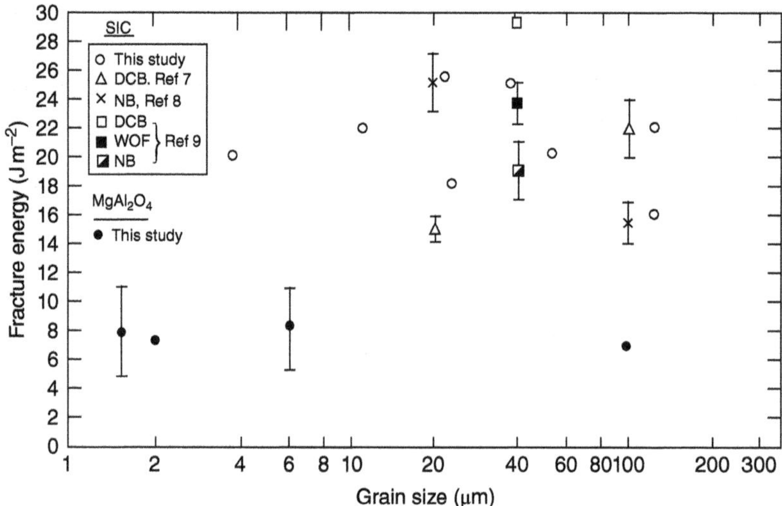

Figure 9.6 Fracture energy as a function of grain size for cubic ceramics. Source: Rice et al. (1981). Reproduced with permission of John Wiley & Sons.

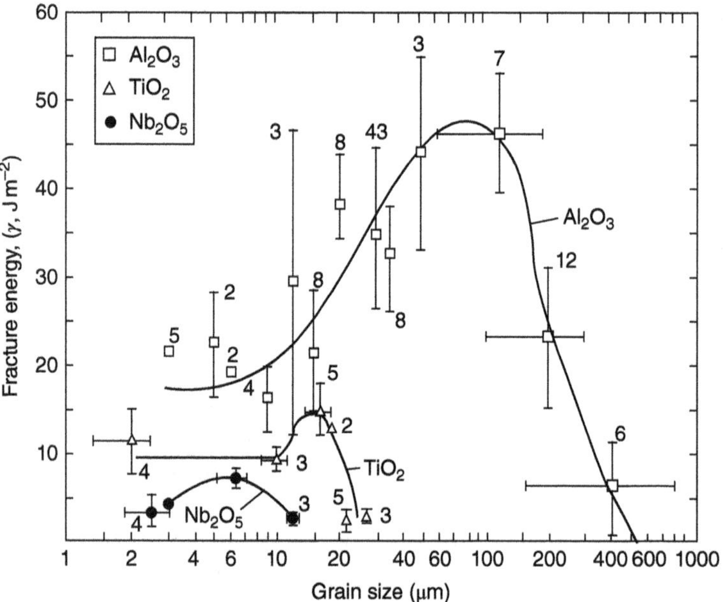

Figure 9.7 Fracture energy as a function of grain size for non-cubic ceramics. Source: Rice et al. (1981). Reproduced with permission of John Wiley & Sons.

not grain size dependent, the strain energy produced by these stresses does increase with increasing grain size that can lead to microcracking. Linking of microcracks causes the significant decrease in K_{IC} at larger grain sizes (Figure 9.7) and can lead to the spontaneous failure of some large-grain ceramics.

While some microcracking ahead of the crack tip has been observed, the more accepted toughening mechanism is a result of stress-induced microcracking, which forms a zone around the crack and shields the crack from the far-field stress. This phenomenon leads to an increase in the load necessary to propagate the crack, hence toughening the material. There are several theories of microcrack toughening. The stress-induced microcrack toughening increment, ΔK_C, can be calculated (Evans 1990):

$$\Delta K_C = \frac{CE\varepsilon^M \sqrt{h}}{1-\nu} \tag{9.4}$$

where E is the elastic modulus, C is a constant that depends on assumptions in the analysis (~0.22), h is the width of the microcracked zone (see Figure 9.8), ε^M is the volumetric strain associated with the microcracking zone, and ν is Poisson's ratio. This toughening can lead to the phenomena of R-curve behavior, i.e. an increase in toughness with an increase in crack size, which is discussed later in this chapter.

Phase Transformations

The transformation of individual grains at a crack tip from one crystal structure to another can lead to significant increases in fracture toughness (Evans 1990). The most common material in which this phenomenon is observed is partially stabilized zirconia, i.e. ZrO_2 in which the tetragonal phase has been stabilized at room temperature through the addition of other constituents, e.g. MgO and Y_2O_3. The picture of toughening in this material is one in which the tetragonal zirconia grains transform to the stable monoclinic structure in the stress field of the crack tip. This transformation produces compressive stresses in a zone surrounding the crack wake similar in appearance to the microcracked zone shown in Figure 9.8. This zone mitigates the crack opening

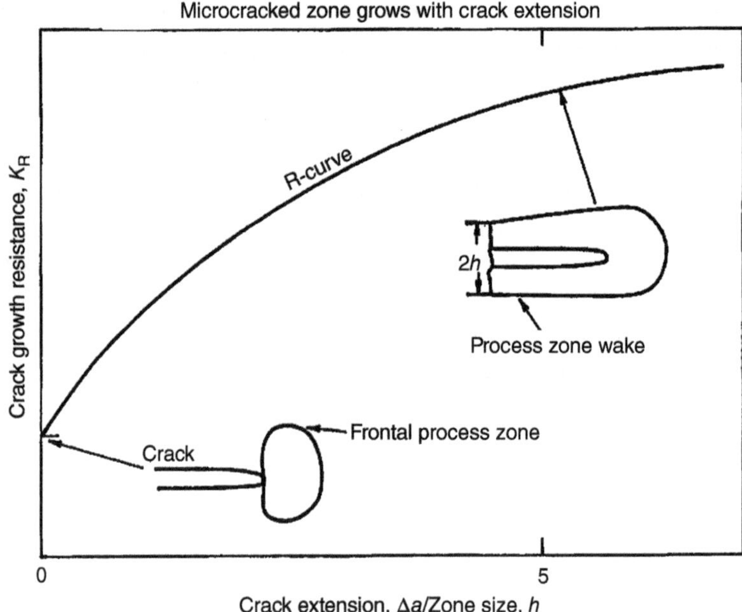

Figure 9.8 Toughening processes leading to a microcracked zone. Source: Adapted from Evans and Faber (1984). Reproduced with permission of John Wiley & Sons.

stresses, thereby partially shielding the crack tip from the high stress intensity.

Fracture toughness values in excess of $10\,\mathrm{MPa \cdot m^{1/2}}$ have been reported (Marshall 1986; Swain and Rose 1986). The K_{IC} of transformation-toughened zirconia is grain size dependent (Marshall et al. 1983). There can also be a trade-off between high toughness and high strength (Swain and Rose 1986). An estimate of the toughness increment, ΔK_C, that can be obtained with transformation toughening can be obtained using an equation analogous to Eq. (9.4) (Evans 1990; Green 1998):

$$\Delta K_C = \frac{CE\varepsilon^T V\sqrt{h}}{1-\nu} \tag{9.5}$$

where E is the elastic modulus, C is a constant that depends on assumptions in the analysis (~0.21–0.41), h is the width of the transformation zone, ε^T is the volumetric strain associated with the transformation, V is the volume fraction of the phase that transforms, and ν is Poisson's ratio.

Another form of phase transformation toughening is observed in ferroelectric materials such as barium titanate. Barium titanate undergoes a transition from a cubic to a tetragonal crystal structure below 150 °C, leading to strains which in turn lead to the formation of 90° twins within the grains. The interaction of these twins with the crack can lead to an increase in K_{IC} of up to 30% (Pohanka et al. 1978).

Crack-surface Tractions and Crack Bridging

There are two primary sources of crack-surface tractions (Figure 9.9), frictional interlocking of grains and bridging of intact material across the crack faces (Freiman and Swanson 1990). These tractions act to shield the crack tip from the applied stress. Since the K_{IC} of such materials depends on the number of traction sites, R-curve behavior can result if the bridging zone increases with crack extension. R-curve materials are those that have increased toughness with increased crack

(i) Frictional interlocking

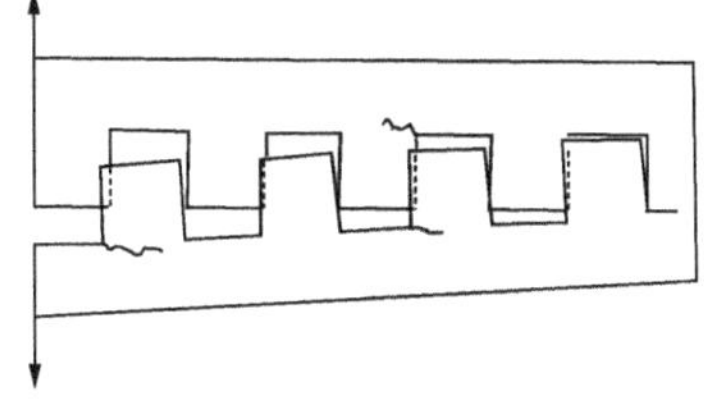

(ii) Ligamentary bridging

(Intact–material "islands" left behind advancing fracture front)

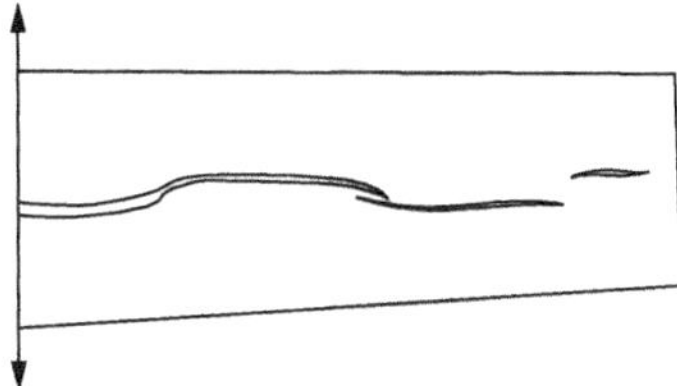

Figure 9.9 Examples of tractions across crack faces that can lead to toughening. Source: From Freiman and Swanson (1990).

size as discussed in the next section. Larger grain size materials and those having elongated grains give rise to greater increases in toughness.

In order to estimate the effect of crack bridging on the toughness of materials, it is necessary to know the details of the stress–displacement relationships for the bridging ligaments. In one particular effect for elastic bridges accompanied by some debonding as the crack faces open, we can estimate the fracture toughness increment, ΔK_{C}:

$$\Delta K_{\mathrm{C}}^2 = V d \sigma_{\mathrm{d}}^2 \tag{9.6}$$

where V is the volume fraction of bridges, d is the debond length, and σ_{d} is the bridge strength (Evans 1990; Green 1998).

R-curves

Materials containing elongated grains, potential transformation phases, stress-induced microcracking, fibers or elongated grains, or multiple grain sizes can exhibit a phenomenon known as a rising R-curve behavior, i.e. an increased resistance to crack growth with increased crack size (Evans 1990). Although the mechanisms around the crack tip can be different, the overall result in a toughness–crack size curve has been similar in shape for all the materials examined that exhibit this behavior, e.g. Figures 9.3, 9.7, and 9.8. Typically, stable crack growth specimens are used to measure the toughness or crack growth resistance parameter as a function of crack size (Nishida et al. 1995). Both compact tension and double cantilever beams (Evans 1974; Yamade et al. 1991) have been used to measure R-curve behavior in ceramic materials. Nishida et al. (1995) suggests a three-point flexure technique using a notched beam with sharpened starter crack. In this technique, it is critical to develop a stabilizing system to insure stable crack growth. They were able to do this by using a system developed by Nojima and Nakai (1993). This is usually performed while monitoring the slowly growing crack with increased load. Slow crack growth can also be accounted for during R-curve measurements (Okada and Hilosaki 1990).

The determination of R-curve values can also be obtained by using the strength of fractured flexure bars, measuring the crack sizes, and

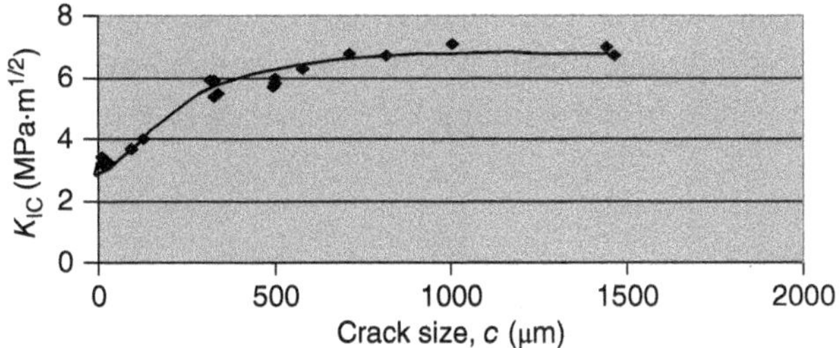

Figure 9.10 Silicon nitride indentation flexure showing R-curve behavior. Source: After Mecholsky et al. (2002).

calculating the toughness. The difference between the determination of R-curve behavior using large cracks versus small cracks has been discussed (Chantikul et al. 1990; Li et al. 1992). If there is no R-curve behavior using small crack specimens such as flexure bars, then the calculated values will be the same for all crack sizes. If there is R-curve behavior, then the calculated values will increase with increased crack sizes up to a certain size. An example of such an R-curve in polycrystalline silicon nitride is shown in Figure 9.10. In this case the mechanism of increased toughness with increasing crack size is grain-to-grain interlocking of the elongated grains in the microstructure, i.e. crack bridging. As the crack encompasses more and more grains, the increased interlocking or bridging across the crack face leads to an increase in fracture toughness.

R-curves have been observed for zirconia, aluminum oxide, and other materials as well. However, increasing values of fracture toughness do not always translate into greater strength.

SUMMARY

In this chapter we have attempted to show the importance of a ceramic material's microstructure, i.e. grain size, shape, and distribution, in determining resistance to flaw growth. We discuss a number of the toughening mechanisms that can lead to increases in fracture toughness and point out that the ratio of the crack size to the grain size is a primary factor in determining crack growth resistance. We also point out that initial flaws can be contained within one large grain, can extend over only a few grains, or can be multiple grains in extent; this factor plays a major role in determining the resistance to the further growth of the flaw.

QUESTIONS

1. Using Figure 9.3, calculate the strengths for ZnSe assuming crack sizes of 100, 500, 1000, and 2000 μm, respectively. Explain the differences in strengths with respect to the toughness, crack sizes, and grain size distribution.
2. What is the implication of (<1 μm) polishing of large-grained materials in terms of strength and toughness?
3. Graph the strength of an R-curve material such as Si_3N_4 (Figure 9.6) as a function of crack size. How does this behavior differ from the behavior of a monolithic amorphous material such as silica glass?
4. If you know that a material has a toughness that behaves as an R-curve material, then which value of toughness would you provide for a structural design of a rotor blade? Why?
5. The measurement of toughness in ZnSe, using a double cantilever beam with a crack that is greater than 1000 μm, is used to determine the strength for cracks that are guaranteed through proof testing to be less than 100 μm. This strength value is greater than the expected applied stress for a given design. What are the problems that may occur in use because of this methodology?

REFERENCES

Chantikul, P., Bennison, S.J., and Lawn, B.R. (1990). Role of grain size in the strength and R-curve properties of alumina. *J. Am. Ceram. Soc.* 73 (8): 2419–2427.

Evans, A.G. (1974). Fracture mechanics determinations. In: *Fracture Mechanics of Ceramics*, vol. I (ed. R.C. Bradt, D.P.H. Hasselman and F.F. Lange), 17–48. New York: Plenum Press.

Evans, A.G. (1990). Perspective on the development of high-toughness ceramics. *J. Am. Ceram. Soc.* 73 (21): 187–205.

Evans, A.G. and Faber, K.T. (1984). Crack growth resistance of microcracking brittle materials. *J. Am. Ceram. Soc.* 67: 255–260.

Faber, K.T. and Evans, A.G. (1983). Crack deflection processes-I. Theory. *Acta Metall.* 31: 565–576.

Freiman, S.W. (2002). Failure from large grains in polycrystalline ceramics: transitions in fracture toughness. In: *Fracture Resistance Testing of Monolithic and Composite Brittle Materials*, ASTM STP 1409 (ed. J.A. Salem, G.D. Quinn and M.G. Jenkins), 17–29. West Conshohocken, PA: ASTM International.

Freiman, S.W. and Swanson, P.L. (1990). Fracture of polycrystalline ceramics. In: *Deformation Processes in Minerals, Ceramics and Rocks* (ed. D.J. Barber and P.G. Meredith), 72–83. London: Unwin Hyman.

Freiman, S.W., Mecholsky, J.J. Jr., Rice, R.W., and Wurst, J.C. (1975). Influence of microstructure on crack propagation in ZnSe. *J. Am. Ceram. Soc.* 58: 406–409.

Green, D.G. (1998). *An Introduction to the Mechanical Properties of Ceramics*. Cambridge, UK: Cambridge University Press.

Hill, T.J., Mecholsky, J.J. Jr., and Anusavice, K.J. (2000). Fractal analysis of toughening behavior in $3BaO{\cdot}5SiO_2$ glass ceramics. *J. Am. Ceram. Soc.* 83 (3): 545.

Kirchner, H.P. and Ragosta, J.M. (1980). Crack growth from small flaws in larger grains in alumina. *J. Am. Ceram. Soc.* 63 (9–10): 490.

Li, C.-W., Lee, D.-J., and Lui, S.-C. (1992). R-curve behavior and strength for in-situ reinforced silicon nitrides with different microstructure. *J. Am Ceram. Soc.* 75 (7): 1777–1785.

Marshall, D.B. (1986). Strength characteristics of transformation-toughened zirconia. *J. Am. Ceram. Soc.* 69 (3): 173–180.

Marshall, D.B., Evans, A.G., and Drory, M. (1983). Transformation toughening in ceramics. In: *Fracture Mechanics of Ceramics*, vol. 6 (ed. R.C. Bradt, A.G. Evans, D.P.H. Hasselman and F.F. Lange), 289–307. New York: Plenum Press.

Mecholsky, J.J. Jr., Hill, R.J., and Chen, Z. (2002). Application of quantitative fractography to the characterization of R-curve behavior. In: *Fracture Resistance Testing of Monolithic and Composite Brittle Materials*, ASTM STP 1409 (ed. J.A. Salem, G.D. Quinn and M.G. Jenkins), 152–168. West Conshohocken, PA: ASTM International.

Nishida, T., Hanaki, Y., Nojima, Y., and Pezzotti, G. (1995). Measurement of rising R-curve behavior in toughened silicon nitride by stable crack propagation in bending. *J. Am. Ceram. Soc.* 78 (11): 3113–3116.

Nojima, T. and Nakai, O. (1993). Stable crack extension of an alumina ceramic in three point bending test (in Jpn.). *J. Soc. Mater. Sci. Jpn.* 42 (475): 412–418.

Okada, A. and Hilosaki, N. (1990). Subcritical crack growth in sintered silicon nitride exhibiting a rising R-curve. *J. Am. Ceram. Soc.* 73 (7): 2095–2096.

Paris, P.C. and Sih, G.C. (1965). Stress analysis of cracks. In: *Fracture Toughness Testing and Its Applications*, ASTM STP 381 (ed. ASTM Committee E-24), 30–81. Philadelphia, PA: ASTM International.

Pohanka, R.C., Freiman, S.W., and Bender, B.A. (1978). Effect of the phase transformation on the fracture behavior of $BaTiO_3$. *J. Am. Ceram. Soc.* 61 (1–2): 72–75.

Rice, R.W., Freiman, S.W., and Mecholsky, J.J. (1980). The dependence of strength-controlling fracture energy on the flaw size to grain size ratio. *J. Am. Ceram. Soc.* 63 (3–4): 129–136.

Rice, R.W., Freiman, S.W., and Becher, P.F. (1981). Grain-size dependence of fracture energy in ceramics: I. Experiment. *J. Am. Ceram. Soc.* 64: 345–350.

Singh, J.P., Virkar, A.V., Shetty, D.K., and Gordon, R.S. (1979). Strength-grain size relations in polycrystalline ceramics. *J. Am. Ceram. Soc.* 62: 179–185.

Swain M. V., Rose, L. R. F. (1986) Strength limitations of transformation-toughened zirconia alloys, *J. Am. Ceram. Soc.* V 69, (7) 511–518.

Yamade, Y., Kawaguchi, Y., Takeda, N., and Kishi, T. (1991). Slow crack growth of mullite ceramics. *J. Ceram. Soc. Jpn.* 99: 467–472.

CHAPTER 10

Reliability and Time-dependent Fracture

BACKGROUND

Two factors are critical in assessing the mechanical reliability of components: the statistical nature of the strength of these materials, as discussed in Chapter 5, and environmentally enhanced flaw growth, leading to time-dependent failure under static stresses (Chapter 3). This chapter provides a description of the test procedures and analysis techniques, including statistical determination of the confidence with which we can accurately combine data from both of these conditions in order to predict the likelihood of failure over a period of time.

TECHNIQUES FOR ASSURING RELIABILITY

Brittle materials contain a range of flaw severities such as size and shape distributed in areas of varying stress. Given this situation, there are three possible approaches to survivability prediction: (i) nondestructive (NDE) evaluation, (ii) proof testing, or (iii) predictions from strength, crack growth data, and statistical analysis.

The Fracture of Brittle Materials: Testing and Analysis, Second Edition.
Stephen W. Freiman and John J. Mecholsky, Jr.

Nondestructive Evaluation

NDE techniques, e.g. dye penetrants, X-rays, and acoustic emission, are used to identify the existence of cracks and other defects in metal parts. The problem with brittle materials is that because of their low fracture toughness, the flaws that lead to failure are of the order of 10–50 μm in depth and most NDE techniques have not been developed to detect those sizes of cracks. Also, the distribution in flaw sizes can be very narrow, meaning that one must identify the existence of a critical flaw that is only slightly larger than the rest of the population. For these reasons, there are currently no known NDE procedures that are effective in determining the safety of advanced, technical ceramic parts.

Overload Proof Test

In concept, the overload proof test is straightforward. One simply applies a stress to each component, typically two to three times the service stress. The ratio of the overload stress to the service stress is the *proof-test ratio*. Any components with critical flaws larger than a predetermined size, or with strengths less than a predetermined minimum (namely the proof-test stress), will break. Accordingly, such components are automatically eliminated from the distribution.

An overload proof test is a procedure for establishing the upper limit on the most critical flaw in a component or, equivalently, *the lower limit on the initial strength distribution*. The proof test is designed to truncate the flaw distribution at a point at which the most severe flaw, and therefore the lowest strength are acceptable in the design of the part; those parts fail during the proof test. The technique is expensive since real components are tested, and some loss of parts must be expected. However, a properly conducted proof test eliminates the problem of failure through statistical outliers. Because the proof test is applied to finished components, it is generally not a technique that is appropriate for material or design selection. In addition, for a final component having a complex geometry, it can be difficult to apply controlled stresses that are equivalent in magnitude and orientation to those experienced in service. See Ritter et al. (1980) for more details.

Strength/Initial Flaw Size Distribution

As mentioned numerous times, fracture in brittle materials usually results from processing-, machining-, finishing-, or handling-induced cracks. The most severe (combination of size, geometry, and orientation) flaw in the component or structure and the spatial distribution of stresses will determine the probability of fracture. The flaw distribution in the component upon its insertion into service is termed the initial flaw distribution, which in turn determines the initial strength. In carrying out a reliability analysis, a key assumption is that no, more severe, flaws are introduced during operation of the component. In addition, it is assumed that the flaws grow only due to environmentally enhanced crack growth.

It would be best if the strength distribution of the actual components could be obtained under loading conditions that simulate the service stresses. This has been accomplished in optical fibers and gas turbine engine components. However, this requirement often poses a difficult engineering problem as well as an economic constraint. Consequently, tests are usually conducted on small pieces of the same material; it is essential that the processing history and surface treatment of these specimens be identical to that in the component.

As pointed out in Chapter 5, because of their simplicity, either uniaxial or biaxial flexural tests are the methods of choice. Because of the larger area interrogated, biaxial tests, e.g. ring-on-ring, are the better choice. Measurements must be conducted at a high loading rate and in an inert environment, e.g. a dry gaseous nitrogen environment, to avoid effects of slow crack growth. If a component is likely to experience a stress state that could cause it to fail from internal flaws such as pores or inclusions, flexural tests will not be effective, and direct tensile tests or their equivalent will be required. The number of specimens that should be tested will depend on the breadth of the strength distribution and the degree of assuredness desired, but 30 is the usually recommended number. In addition, failure sources should be identified through fractography to be able to separate those resulting from processing defects, e.g. pores, from those caused by machining or handling. In Chapter 5, we showed that based upon these laboratory strength data, one can calculate the minimum strength of a set of components with a specified degree of confidence, i.e. the lower tolerance limit (LTL).

Determining Environmental Susceptibility

The detailed mechanisms of crack growth are best determined through direct observation of a growing crack (Chapter 3). However, for a survivability analysis, tests employing cracks that are of comparable size with actual flaws in the component are both more convenient, as well as more relevant to the practical failure issue, i.e. strength tests.

Time-to-failure Expressions

Our ability to predict the reliability or lifetime of a ceramic part whose strength is likely to degrade with time rests on the use of a mathematical expression that can represent the crack growth data.

As shown in Chapter 3, the following expression has been shown to explain environmentally enhanced crack growth curves:

$$V = V_0 \exp(bK_{\mathrm{I}}) \tag{10.1}$$

where V_0 is a material-dependent parameter containing terms such as the relative humidity of water and the temperature and b is a parameter proportional to the activation volume for the process. It is useful to put the stress intensity factor, K_{I}, in terms of the stress, σ, in the component through

$$K_{\mathrm{I}} = Y\sigma a^{\frac{1}{2}} \tag{10.2}$$

where a is the flaw size, taken to be the depth of a surface crack, and Y is a geometric constant dependent on the type of loading, the flaw geometry, and location. σ is the far-field stress in the specimen or part.

Michalske et al. (1991) showed that by combining Eqs. (10.2) and (10.1), and integrating over crack length, one can obtain the time to failure as a complex function of the critical and initial flaw sizes. Because their integration limits are in terms of flaw size, a difficult measurement to make, the prediction will likely be inaccurate. In addition, the effect of a residual stress term associated with the introduction of the flaw has not been incorporated into such an expression (Fuller et al. 1983).

An empirical power law is used instead. A comparison of the fits to data of the two forms has been reported (Jakus et al. 1981), and over a large range of times agree reasonably well:

$$V = V_0 \left[\frac{K_I}{K_{IC}} \right]^N \tag{10.3}$$

where V_0 and N are material-dependent constants. N ranges from as low as 10 for glasses to 50 or greater for ceramics such as silicon nitride and silicon carbide. K_{IC} is the critical stress intensity factor, which in the present context is a scaling parameter.

Because failure in brittle materials typically begins from surface flaws caused by grinding, machining, or some form of handling, it is reasonable to assume that there will be a local residual stress associated with each flaw generated by the way they were introduced. Equation (10.4) is modified to the following (Fuller et al. 1983):

$$V = V_0 \left[\frac{K_I}{K_{IC}} \right]^{N'} \tag{10.4}$$

where $N' = (3N+2)/4$ for point-type flaws, i.e. small, semielliptical surface cracks.

We also use the fact that

$$K_I = Y\sigma a^{\frac{1}{2}} \tag{10.2}$$

σ is the far-field stress on the flaw, a is the semiminor axis of an elliptical surface crack, and Y is a geometric constant dependent on flaw geometry and location and is a measure of the flaw severity.

Because of the nature of the indentation-induced cracks, Fuller et al. (1994) derived an expression for time to fracture that involves only strength measurements:

$$t_f = \frac{\lambda}{N'+1} \left(\frac{S}{S_v} \right)^{N'-2} \sigma^{-N'} \tag{10.5}$$

where S_v is the inert strength of an indented specimen of the material. λ and N' are determined through measuring strength as a function of stressing rate, i.e. what are termed "dynamic fatigue" tests (ASTM C1368-06 2017):

$$\sigma_f^{N'-1} = \lambda \frac{d\sigma}{dt} \tag{10.6}$$

In dynamic fatigue testing, an indented specimen is subjected to a load that increases linearly with time (Figure 10.1). By employing indentations resulting in cracks that are surrounded by a stress field that includes tensile stresses, one is assured that a more conservative value of N is obtained, i.e. N' rather than N. A linear regression of the log of the failure strength, σ_f, on the log of the stressing rate, $\dot{\sigma}$, gives the parameters N' and λ. These tests should be conducted in the harshest environment that is likely to be seen by the structure or component, usually liquid water.

It is important to obtain the dynamic fatigue data over as wide a range of stressing rates as possible, e.g. 4–5 orders of magnitude. *Data taken over less than 3 orders of magnitude in stressing rates should not be trusted.* ASTM Standard C1368-10 provides detailed instructions on the dynamic fatigue test. Guidelines for the stressing rate limits on testing machines and load cells should also be consulted.

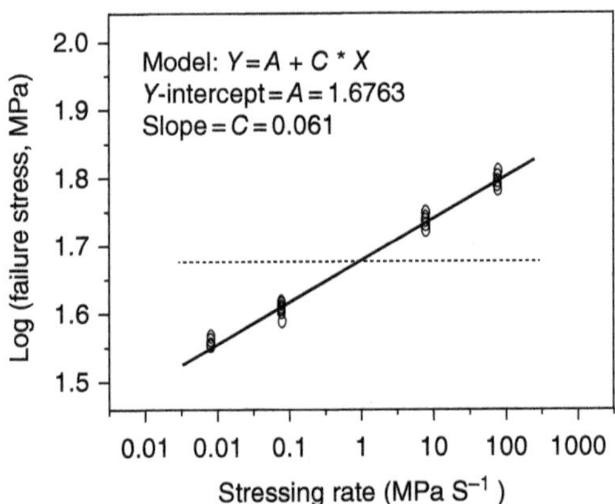

Figure 10.1 Dynamic fatigue data for a borosilicate glass (BK7) tested in water.

If indented specimens were used to determine N', measurements are needed to calculate S_v, the upper limit to the indented strength. Again, these strength measurements must be made at a high loading rate in an inert (e.g. N_2 gas) environment.

Service Stresses

The stress state in the component is typically determined using a finite element analysis. While there is a tendency to accept that a finite element analysis is without error, inexact knowledge of parameters such as the elastic modulus can lead to uncertainty in the stress prediction.

Uncertainty Calculations

It is not sufficient from the point of view of safety to simply calculate a time to failure; one must also know the uncertainty in this calculation, particularly the lower bound. There is a three-step strategy to estimate the uncertainty of the failure time in the full-scale structure.

Step 1

The first step is to determine the minimum strength of the specimens, i.e. the LTL as described in Chapter 5. To determine the uncertainty in the minimum S, one fits the strength distribution with a function, e.g. the three-parameter Weibull function, and calculates the breadth of the uncertainty using established statistical techniques (Fong et al. 2018). Goodness-of-fit tests (Bury 1975) are used to establish the best statistical fit to the data. One can then choose the acceptable level of confidence desired for the lower bound in this calculation.

Step 2

From Eq. (10.8), we can see that the time to failure, t_f, is a function of five parameters, namely, λ, N', S, Sv, and σ. A formula for the variance of the time to failure, var (t_f), can be derived using established error propagation formulas as a function of the variances of those five parameters (Ku 1966). The details of this derivation are given in

Fong et al. (2018). The expression for the standard deviation (sd) of t_f is given by

$$\mathrm{sd}(t_f)=t_f\sqrt{\frac{\mathrm{var}(\lambda)}{\lambda^2}+\frac{\mathrm{var}(N')}{(N'+1)^2}+(N'-2)^2*\left[\frac{\mathrm{var}(S)}{S^2}+\frac{\mathrm{var}(S_v)}{(S_v)^2}\right]+(N')^2\frac{\mathrm{var}(\sigma)}{\sigma^2}} \quad (10.7)$$

Note that the standard deviation in t_f increases in direct proportion to the time to failure.

We can derive two simplified formulas from Eq. (10.8).

CASE 1

Assuming var(λ), var(N'), and var(S_v) are negligible as compared with the var(S), and that the var(σ) is also negligibly small, we obtain

$$\mathrm{sd}(t_f)=(N'-2)*\left(\frac{t_f}{S}\right)*\mathrm{sd}(S). \quad (10.8)$$

CASE 2

If there is uncertainty regarding the finite element analysis, then

$$\mathrm{sd}(t_f)=t_f\sqrt{(N'-2)^2\frac{\mathrm{var}(S)}{S^2}+(N')^2\frac{\mathrm{var}(\sigma)}{\sigma^2}} \quad (10.9)$$

If Eq. (10.9) yields uncertainties in failure times that are unacceptably large, there are three options to consider. Whereas the parameters N', λ, and S_v are functions of the material selected, and therefore would not be subject to change, it may be possible to increase the strength of the material and thereby shift the position of the Weibull distribution as well as increase λ by better machining and polishing procedures. Second, it may be possible to reduce the stress state in the component through a different mechanical design. A third, and simpler, option may be to increase the size of the initial strength data set. The more specimens that are tested the better is the knowledge of the entire flaw size distribution and therefore the smaller the variance.

Step 3

The preceding uncertainty analysis was based on laboratory data taken on uniaxial or biaxial flexural specimens. We must then translate this data to provide the probability of failure within a set of components with which the minimum lifetime of the component can be predicted. The calculation of coverage allows one to provide safety assurance over as wide a percentage of the entire number of components as needed depending on the degree of safety needed.

SUMMARY

This chapter describes procedures for determining the reliability of ceramic components under static stresses. The chapter focuses on statistical methods of estimating uncertainties in the measured fracture parameters. It is shown that error propagation models can be used to combine uncertainties in laboratory test data and thereby determine a standard deviation for a predicted time to failure. The reliability of

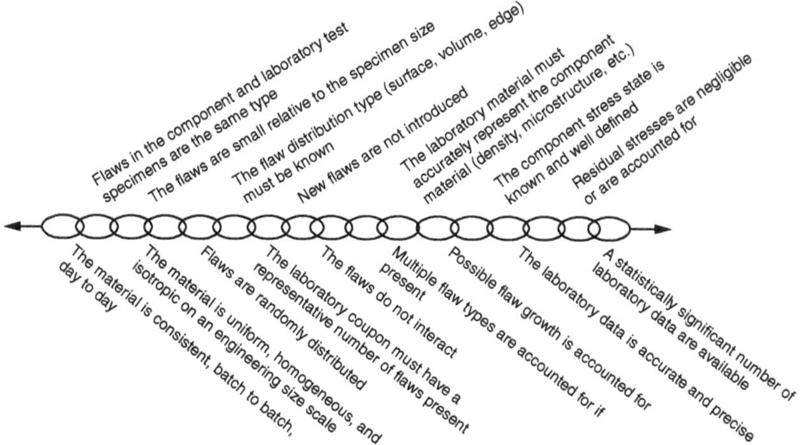

Figure 10.2 Factors influencing the accuracy of component strengths from laboratory data. Source: Quinn (2005). Reproduced with permission of John Wiley & Sons.

actual components is obtained from the data obtained on specimens through the concepts of "tolerance limit" and "coverage."

We should point out that the accurate prediction of component strengths based on laboratory flexural strength measurements is dependent on many factors, as was pointed out by Quinn (2005). See Figure 10.2 for a schematic illustration of these issues.

Finally, one should recognize that the approach described herein can be used to predict failure caused other than that due to environmentally enhanced crack growth. The requirement is a relationship between predicted time-to-failure and various measurable parameters.

QUESTIONS

1. For a material that is susceptible to enhanced crack growth due to the environment, will the strength be greater or less with increasing stressing rate? Why?
2. Describe the proof-test procedure. How can you insure that crack growth does not occur during loading and unloading? How would the strength distribution change if there was crack growth during loading?
3. How would you determine if a set of flexure bars that were ground had residual stress associated with the cracks induced by the grinding? How could you eliminate the residual stress if necessary? How would the strength as a function of stressing rate test procedures change if residual stress is present or if it is not present?
4. Derive Eq. (10.6) assuming Eq. (10.2) governs environmentally enhanced crack growth.

REFERENCES

ASTM C1368-06 (2017). *Determination of Slow Crack Growth Parameters of Advanced Ceramics by Constant Stress-Rate Flexural Testing at Ambient Temperature*. West Conshohocken, PA: ASTM International.

Bury, K.V. (1975). *Statistical Models in Applied Science*. New York: Wiley.

Fong, J.T., Filliben, J.J., Heckert, N.A. et al. (2018). Estimating with uncertainty quantification a minimum design allowable strength for a full-scale component or structure of engineering materials. *J. NIST Res.* (To be published).

Freiman, S.W., Fong, J.T., Heckert, N.A., and Filliben, J. (n.d.). New statistical methodology for assessing mechanical reliability: II, environmental crack growth. (To be published).

Fuller, E.R. Jr., Lawn, B.R., and Cook, R.F. (1983). Theory of fatigue for brittle flaws originating from residual stress concentrations. *J. Am. Ceram. Soc.* 66: 314–321.

Fuller, E.R. Jr., Freiman, S.W., Quinn, J.B. et al. (1994) Fracture mechanics approach to the design of glass aircraft windows: a case study. *SPIE Conference Proceedings – The International Society for Optical Engineering*, San Diego, CA (26–28 July 1994), Vol. 2286, pp. 419–430.

Jakus, K., Ritter, J.E., and Sullivan, J.M. (1981). Dependency of fatigue predictions on the form of the crack velocity equation. *J. Am. Ceram. Soc.* 64: 372–374.

Ku, H.H. (1966). Notes on the use of propagation of error formulas. *NBS J. Res.* 70C (4): 263–273.

Michalske, T.A., Smith, W.L., and Bunker, B.C. (1991). Fatigue mechanisms in high-strength silica-glass fibers. *J. Am. Ceram. Soc.* 74: 1993–1996.

Quinn, G.D. (2005). Design and reliability of ceramics: do modelers, designers, and fractographers see the same world. *Ceram. Eng. Sci. Proc.* 26: 239–252.

Ritter, J.E. Jr., Oates, P.B., Fuller, E.R. Jr., and Wiederhorn, S.M. (1980). Proof testing of ceramics, part 1, experiment. *J. Mater. Sci.* 15: 2275–2281.

CHAPTER 11

Concluding Remarks

In this, the second edition of our book, we have presented the background, testing procedures, and analysis methodology needed to assess the fracture behavior of brittle materials, particularly ceramics. We emphasize that ceramics should be considered as a broad category of brittle materials, ranging from single crystals of semiconducting compounds to polycrystalline bodies containing multiple grains as well as grain boundary phases. Although we focus on the fracture of ceramics, most of the principles and test methodology are applicable to any material that fails in a brittle manner, including metals and polymers. In this edition of the book we expand our discussion of materials to include biomaterials. In addition, two new chapters have been added. One, Chapter 6 discusses the fracture of brittle materials subjected to thermally induced stresses. Chapter 7 describes efforts to calculate fracture toughness based upon the atomistics of the material.

The selection of a test to determine fracture parameters is dictated by a number of factors, e.g. the type of material, its microstructure, the geometry of the final part, and how the data will be used. For example, if the material is amorphous, it should not matter what size of crack is

The Fracture of Brittle Materials: Testing and Analysis, Second Edition.
Stephen W. Freiman and John J. Mecholsky, Jr.

employed to obtain the fracture properties. As discussed in Chapter 9, the fracture behavior of polycrystalline materials could be dependent on the size of the crack relative to the size of the microstructure. In addition, if there are microstructurally related toughening mechanisms available for the material, then fracture toughness will also depend on flaw size, giving rise to *R*-curves. The fracture resistance of single crystals will depend on crystal orientation, as will that of thin films and coatings. In addition, the form of the final product will also influence test selection. Materials that are manufactured in the form of rods or cylinders will demand a different test than those formed as sheets or plates. Chapter 5 describes various strength test procedures available depending on the form of the material and its final use.

An important consideration is how the data will be used. Fracture toughness data is typically used to rank materials for initial selection, although it can be used more directly as a measure of a material's resistance to impact and wear. It is not typically used directly in the prediction of strength or lifetime. As we have tried to emphasize in Chapter 9, because of the effect of the ratio of crack size to grain size in determining resistance to fracture, higher fracture toughness values may not necessarily lead to higher strengths. As we note, initial flaws can be small enough to reside within one grain or encompass only a few grains, leading to strengths much lower than would have been predicted from macroscopic fracture mechanics tests. It is incumbent upon the designer or practitioner to understand the effect of microstructure on the process. The distribution of flaw sizes as well as the distribution of grain sizes could affect the final results if the values placed in the program are not representative of the material because the relationship of the microstructure to the crack sizes was not accounted for.

Most materials that fail in a brittle manner are subjected to a susceptibility to slow crack growth due to environmental conditions, as discussed in Chapter 3. This phenomenon is critical to behavior and design. It means that a material may survive initial loading, but may fail after some time under load. When determining the effects of slow crack growth, the research or designer must perform strength (or fracture toughness) tests in the same type of environment as expected in service.

We stress the importance of knowledge of the source of failure through fractographic examinations. All materials that fail in a brittle manner have similar fracture characteristics. That is, the material's fracture process usually starts from one primary source. Knowledge of the kind of critical flaws, their location, shape, etc. is important to be able to design fail-safe components. When fracture initiates, the crack path radiates outwardly in a relatively symmetric pattern. Most of the time, there are characteristic regions surrounding the origin, known as mirror, mist, and hackle, which are separated by increased steps in roughness. Finally, the crack separates into two or more crack directions. The details of the process are dictated by the amount of energy available to fracture the material and can be described using the principles of fracture mechanics. Fracture surface measurements provide free information, i.e. it exists as a result of the fracture process. Quantitative fractography, as described in Chapter 8, is an important adjunct to an overall understanding of the material's behavior during a fracture event.

We cannot emphasize too strongly the importance of the statistical nature of brittle fracture. Knowledge and proper analysis of the variation in strength values emanating from a distribution in flaw severities is vital in the design of ceramic structures and components that can withstand operational stresses over long periods of time. This puts a stringent requirement on the number and kind of specimens necessary to obtain sufficiently accurate values of the strength distribution. A further important issue is that component sizes can be significantly larger than the test specimens, meaning that the area or volume under stress is proportionally larger and the strength lower. Chapters 5 and 10 contain suggested statistical techniques to analyze these variabilities.

Predictions of the mechanical reliability of a component are based on a combination of all the elements discussed in this book. The designer needs reliable values for the toughness, a good estimate of the expected strength distributions, and knowledge of the potential for slow crack growth due to environmental conditions. All of these elements are required for a good analysis in lifetime predictions. Fortunately there are computer programs available to aid the designer. The procedures for predicting the lifetime of components subjected to stresses in a given environment are outlined in Chapter 10.

Finally, throughout, we have attempted to provide the reader with references to the original studies. We encourage the examination of these publications to obtain many of the details of the fracture behavior of these materials.

Subject Index

Activation energy, 26
Activation volume, 25
Aluminosilicate glass, 28
Aluminum oxide, 41, 143, 219
  fracture energy of, 151, 161
  quenching of, 139
Anisotropy, 111, 152, 169, 177, 193, 213
Applied moment DCB, 40, 41
Area
  fracture surface, 212
  of loading curve, 58
  slit island, 195, 197
  stressed, 8, 116, 127, 155, 156, 160
ASTM Standards
  C1161-13, 119
  C1211-18, 119
  C1239-07, 110, 113, 119, 191
  C1273-15, 119
  C1322-96a, 177
  C1323-16, 95, 98, 119
  C1368-06, 228
  C1421-01b, 53
  C1499-01, 80–82, 84, 96, 97, 120, 165
  C1525-04, 135
  C1684-18, 119
Atomic
  AFM, 54, 111, 125, 128
  bonds, 8, 12, 128
  dimension, 8, 125, 128
  force microscope, 179, 195
  orbitals, 30

Ball-on-ring test, 84, 87, 165
Ball-on-three-balls test, 84, 89
Barium titanate, 3, 217
Biaxial flexure tests, 80, 84, 85, 105, 106, 120, 124
Biaxial stress, 86, 100, 122
Biolox, 3
Biomaterials, 5, 103–105, 168
Biot number, 132
Bond energy, 150
Borosilicate glass, 28
Branching angle, 189, 190
Brazilian disk test, 100
Brittle fracture, failure, 3–8, 108, 160, 170, 223, 229
Brittleness, 1, 2, 81, 168

Ceramics, 1–4, 32, 38, 40, 44, 207, 213–215, 218, 219
Chemical reaction rate theory, 26
Chevron notch bend test, 53–55
Circumferential tension, 87, 97–99, 123
Cleavage, 146
Compliance, 13, 87
Compression, 96–98, 100–103, 163

The Fracture of Brittle Materials: Testing and Analysis, Second Edition.
Stephen W. Freiman and John J. Mecholsky, Jr.

Compressive strength, 103
Contoured DCB specimen, 42
Coplanar maximum strain energy release-rate criterion, 15
Coverage, 116–118
Crack branching, 171–173, 193, 198, 200
Crack bridging, 212, 218, 219
Crack extension, 24, 58, 68, 210, 217
Crack growth rate, 26–35, 45, 49, 145
Crack growth threshold, 29
Crack tip, 1, 7, 9–12, 16, 17, 26–35, 39, 40, 45, 176, 191
  process zone at, 215–218
C ring specimen, 95–98, 110–112, 121
Critical stress intensity factor, 12, 64, 100, 127, 130, 146

DCDC Specimen, 59
Defect(s), 2, 4, 7, 19, 32, 72, 90, 207
Delayed failure, 23–25, 116–118
Density functional theory, 152
Dental materials, 1, 2, 104, 188
Diametral compression test, 62, 63, 100, 101
Double cantilever beam specimen, 38–49
  applied moment, 40–45
  contoured, 42
  wedge loaded, 43
Double cleavage drilled compression specimen, 53, 54
Double torsion specimen, 47, 57, 73
Dynamic fatigue, 228

Edge chipping, 66–68
Elastic modulus, 9, 146, 147, 149, 215, 216
Elliptical integral, 12
Environmentally enhanced crack growth, 13, 23–36, 54, 65, 223, 225, 226, 232
Equibiaxial strength, 84
Error propagation, 231, 232

Failure criterion, 148
Failure probability, 111, 112
Fiber tests, 94, 95
Finishing operation, 2, 102, 151
Finite element analysis, 12, 94, 98, 102–105
Flaws, 3, 8, 17, 79, 113, 114, 122, 207–211, 219
Flexural testing
  biaxial, 80, 84, 85, 105, 106, 120, 124, 225, 231
  2-point bend, 95
  3-point flexure, 50, 80, 82, 105, 113, 121, 134, 142
  4-point flexure, 80, 82, 83, 119, 130
  uniaxial, 80, 225, 231
Fractal analysis
  dimensions, 156, 159, 194, 195, 197–199
  geometry, 156, 197
Fractography, fractographic analysis, 5, 51, 83, 89, 94, 106, 107, 109, 111–114, 118, 122, 123, 128, 129, 152, 159, 167–169, 171, 175, 177–190
Fracture
  atomistics of, 162
  fast, 82
  indentation, 60, 61
  mirror, 15, 82, 101, 109, 111–120, 125, 128–130, 159, 167–170
  mixed mode, 9, 13–17, 60–64
  slow, 19–31
Fracture energy, 9, 12, 14, 146, 149, 151, 210
Fracture mechanics, 10, 15, 20, 32, 52, 60, 65, 66, 99, 100, 127–129, 134

Fracture surface, 8, 10, 15, 27, 51, 105, 158
Fracture toughness, 12–14, 17, 37–74

Gallium arsenide, 2
Generalized maximum tangential stress criterion, 18, 19
Glass, 2, 7, 8, 23–35, 42, 45, 49, 56, 64–67, 84
Glass fibers, 2, 8
Goodness of fit, 229
Grains, 207–220
Griffith theory, 7, 8, 12–18
Groove, grooving, 39, 42, 44, 45, 47, 55, 73

Hackle, 170–172, 177, 179, 187, 189–195, 200, 201, 237
Hardness, 2, 61, 62, 64, 66, 68
Hardness impression, 55, 60, 61
Hartree Fock model, 154
High temperature testing, 61, 92, 93, 100
Hourglass specimens, 104, 105

Indentation cracks, 68, 90, 112, 114, 121, 163
Indentation fracture technique, 60–62
Intrinsic origins, 169
ISO Standards, 48–50
  ISO, 14704:2016 (flexural strength, RT), 120
  ISO, 15490:2008 (tensile strength, RT), 120
  ISO, 15732:2003 (flexural strength, SEPB), 56
  ISO, 6872 (Dental Ceramics), 89, 110
  ISO, 17565:2003 (flexural strength, HT), 120
  ISO, 18756:2003 (flexural strength, SCF), 56
  ISO, 23146:2008 (flexural Strength, SEVNB), 56
  ISO, 24370 (fracture toughness, CNB), 54

K_{Bj} 16, 104, 113, 116
K_{IC}, 11–14, 32, 34, 36, 39, 40, 43, 45, 48–51, 55–57, 65, 66, 209, 219
Kinks in crack, 155
Knoop indentation, 49

Laminate testing, 105–108
Lattice trapping, 152
Lifetime prediction, 237
Loading modes, 10
Lower confidence bound, 116
Lower prediction bound, 116
Lower tolerance limit, 225, 229

Machining, 2, 16, 17, 43, 52, 53, 83, 90, 92–95, 112–114
Maximum normal stress criterion, 14, 16
Maximum strain energy release-rate criterion, 14, 16, 64
Mechanisms of crack growth, 30–35
MEMS devices, 101
Metallic glasses, 150
Microcracking, 213–217
Military handbook, 177
Mirror, 15, 94, 170–199
Mirror-to-crack size ratio, 117, 128
Mist, 15, 170–200
Mixed mode, 14–19, 68–72, 101, 109, 119, 135, 136, 211
Molecular dynamics calculations, 155, 160
Molecular orbital calculations, 153

NIST Recommended Practice Guide, 109, 119, 177, 187, 190
Non-coplanar maximum strain energy release-rate criterion, 14
Nondestructive evaluation, 224

Optical fiber, 102, 105, 166, 168, 169
O ring test, 95–99, 110–112

Partial pressure of water, 12, 27
pH, 27
Phase transformation(s), 213, 215, 217
Plane strain, 12–14, 42, 100, 163, 169
Plane stress, 12, 13, 100
Plastic deformation, 7
Polycrystalline, 39–41, 45, 65, 66, 95, 119, 120, 127, 209, 210, 212, 215, 219
Pores, 169
Probability of failure, 91–95, 122, 123, 148, 155, 190, 191
Process zone, 216
Proof test, 94, 223, 224

Quartz, fracture energy of, 151
Quenching, 135

Radial cracks, 60, 163
R-curve, 215, 217–219
Regions of crack growth, 25, 28, 29, 32, 116
Relative humidity, 12, 45
Reliability, 2, 4, 5, 223, 225–227, 229
Residual stress
  global, 129
  local, 129, 137, 150, 169, 182
Retained strength, 140
Ring testing
  ball-on-ring test, 84, 87, 165
  piston-on-ring test, 84, 85, 89
  ring-on-ring, 84–87, 114, 225
  rings in flexure, 95

Scaling, 194, 195, 198
Silica, 28, 144, 154, 159, 161
Silicon, 2, 152, 154, 161
Silicon carbide, 2, 94, 146, 151, 227
Silicon nitride, 2, 3, 227
Single edge notched beam test, 53
Single edge pre-cracked beam test, 55
Slow crack growth, 13, 16, 61, 99, 113, 116, 117, 129, 130, 135, 142, 148, 151, 156, 159, 160, 164, 165, 171
Soda-lime-silica glass, 24, 26, 27, 95, 108, 113, 116–118, 121, 125, 149, 188, 192
Sources of failure, 71, 94, 101
Stable crack growth, 50, 57, 62, 82, 142
Static fatigue, 19, 23
Statistics of fracture, 108–113, 156, 223, 229
Strain energy release rate, 10, 15, 17, 18
Strength
  strength, indentation, 61
  strength distribution, 109–115, 122, 223, 225, 231
Stress corrosion, 25
Stress distribution, 83, 87, 89, 90, 98, 104
Stress intensity factor, 11, 24–35, 37, 48, 54, 57, 60, 62, 71, 72, 134, 191, 193, 197, 211, 216
Subcritical crack growth, 11, 13, 63, 157, 162, 164, 171
Surface crack, 11–14, 52, 56, 64, 80, 109
Surface crack in flexure test, 56

Tapered DCB specimen, 42
Tensile tests, 79, 90–95, 104, 114, 152
Thermal expansion anisotropy, 101, 213
Thermal shock, 5, 130
Thermal shock tests, 135–141
Thermal stress, 130–133
Theta specimen, 101, 102, 120
Time to failure, 116
Tolerance limit, 155, 156
Toughening, 207, 212, 213, 215–217, 219
Toughness tests
  chevron notch bend test, 48, 49

double cantilever beam, 32–38
double cleavage drilled compression (DCDC), 53, 54
double torsion, 40–43
single edge notched beam, 47
single edge pre-cracked beam, 49, 50
Tractions, 217
Transducers, 1
Tubular components, 98–100, 110
Twist hackle, 177, 179, 187, 191–193

Uncertainty calculations, 109, 113, 115, 116, 118
Uniaxial tension, 92, 94, 119
Universal Binding Energy Relation (UBER), 151

Variance, 229, 231
Vickers hardness indentation, 39, 49, 50, 55, 60, 61, 164
Vitreous silica, 29, 33
Volume defects, 17, 102, 103

Wake hackle, 187, 189
Wallner lines, 187, 188, 190
Wedge loaded DCB specimen, 38, 39
Weibull distribution, 109–116, 153–156, 229
Work of fracture test, 58

Zinc Selenide (ZnSe), 177, 179
Zinc Sulfide (ZnS), 161
Zirconia (ZrO_2), 57, 107, 120, 141, 143